Exploring
CALCULUS

WITH A GRAPHING CALCULATOR

Exploring

CALCULUS

WITH A GRAPHING CALCULATOR

CHARLENE E. BECKMANN
THEODORE A. SUNDSTROM

Grand Valley State University

ADDISON-WESLEY PUBLISHING COMPANY
Reading, Massachusetts • Menlo Park, California • New York
Don Mills, Ontario • Wokingham, England • Amsterdam • Bonn
Sydney • Singapore • Tokyo • Madrid • San Juan • Milan • Paris

ISBN 0-201-55574-3

6 7 8 9 10-CRS-97 96 95 94

Preface

Exploring Calculus with a Graphing Calculator is just that, exploratory. Using graphs and numbers, an understanding of the ideas of calculus is built first on familiar applications. Applications are modeled graphically and then numerically, helping students build an understanding of calculus concepts through graphs and numbers. *Exploring Calculus* has been piloted and revised as a result of reactions from calculus students over the past several years.

Exploring Calculus has been written generally enough to use with many graphics tools. To make best use of the explorations, a graphics tool and a numeric tool (preferably programmable) for generating tables of values will be necessary. The graphics tool should have the following capabilities:

- It should be possible to easily change the domain and range to graph a function on any interval.
- It should be capable of graphing the following functions: polynomial, sine, cosine, tangent, natural log, natural exponential, absolute value, greatest integer, hyperbolic sine and cosine, inverse sine, inverse cosine, and inverse tangent.
- It should have a trace option that will allow values to be read right off the graph.
- It should have zoom-in capabilities; however, this can be accomplished through resetting the domain and range, or through the use of an additional program (included for CASIO graphics calculator).

In addition, some of the materials were written to be used with particular programs on a computer or graphics calculator. Programs for the CASIO and Texas Instrument Graphics Calculators are included here in Appendices A and B, respectively. Similar programs are available in various calculus software packages for the IBM, Apple II series, and Macintosh computers. Public domain software is also available from the authors. While most of the materials can be used without such programs, it will be helpful to obtain a computer package with programs to do the following:

- Sketch secants along the curve of $y = f(x)$,
- Sketch tangents along the curve of $y = f(x)$,
- Use Newton's Method for finding a root, both graphically and numerically,
- Graph left and right endpoint approximations for Rieman sums,
- Approximate integration routines for midpoint, trapezoid, and Simpson's rules,
- Graph parametric curves, and
- Sketch two-dimensional vectors.

Foreword for CASIO and TI Graphics Calculator Users

While *Exploring Calculus* has been written generally enough to use with many graphics tools, specific directions for use of the Texas Instrument TI-81 graphics calculator or any of the CASIO graphics calculators, fx-7000G, fx-7500G, fx-8000G, fx-8500G, or fx-7700G, have been included. Use of any of these graphics calculators is suitable. However, to avoid having to clear programs from the calculator memory, it might be best to obtain a machine with at least 1000 bytes of memory.

For most efficient use of the materials, specific suggestions for use of the CASIO and TI graphics calculators are listed in footnotes. A separate appendix for each of these tools is included. Each appendix includes the following:

- A list of the materials and the programs used to complete the particular calculus exploration;
- An introduction describing operations and programming conventions of the graphics calculator;
- A reference sheet with commonly used commands for the graphics calculator;
- The Diving Board Problem, a precalculus problem written to help familiarize users with the graphic and numeric capabilities of the calculator, with specific directions for operation of the graphics calculator in both interactive and programming modes and complete directions for entering and running a short program;
- Programs used in the materials, listed in alphabetical order, including a description of the purpose of the program, the code and function of each line of code, and a sample run.

Students will find that reading the introduction and using the reference sheet will save time and minimize frustration later. Students who have limited familiarity with the graphics calculator will find it helpful to work through the Diving Board Problem. This precalculus problem includes specific instructions, including keystrokes, to familiarize the student with the graphic and numeric capabilities of the calculator. Also, detailed directions for entering and running a program are provided.

Acknowledgements

This project continues to evolve. *Exploring Calculus* has been piloted, modified, and reused as a result of reactions from students, colleagues, and friends over the past several years. The authors wish to acknowledge the contributions of the countless students who have been so candid in their comments and suggestions. In particular, comments and suggestions from students at Grand Valley State University and Muskegon Catholic Central High School have been very helpful.

Colleagues have made important suggestions for improvements. The authors are especially indebted to Tom Gruszka (Grand Valley State University), who spent a considerable amount of time reading and reacting to several of the explorations. Thanks also to Christi Bruns (West Ottawa High School), Kathy Heid (Penn State University), Diane Krasnewich (Muskegon Community College), Cameron Nichols (Kalamazoo Area Math/Science Center), and Deborah Schaalma (Muskegon Catholic Central High School) for sharing ideas and making suggestions as the first and then second editions were being prepared. The authors also wish to thank Charles Vonder Embse (Central Michigan University) for allowing the use of the program ZOOM (CASIO).

Dedication

This work is dedicated to Ben, Colleen, Melanie, and Dave, for their patience and support throughout the writing of *Exploring Calculus* and particularly in the last several months while this work was being completed.

Table of Contents

xii

Exploring

CALCULUS

WITH A GRAPHING CALCULATOR

1

Exploring Families
of Functions

1.1 Introduction

The objective of this investigation is to determine the effect on the graph of $y = f(x)$ of changing the parameters A, B, C, and/or D in the function $g(x) = A \cdot f(Bx + C) + D$. How does the new function g relate to the original function f when one or more of the parameters A, B, C, or D is changed? In particular, how are the domain, range, and symmetries of a function changed when the function f is changed as in g? How does the shape of the graph of the function change? Most importantly, what effect does each of these parameters have on a function f that causes these changes to occur?

1.2 Exploration

1. a.   Graph each of the following functions, with appropriate domains and ranges. Record the results with paper and pencil, labelling the graphs and including the scales on both axes.

function	$y = f(x)$	f is the parent function of:
1.	$y = x$	$y = mx + b$
2.	$y = x^2$	$y = ax^2 + bx + c$
3.	$y = |x|$	$y = a\,|bx + c| + d$
4.	$y = \sin x$	$y = a \sin (bx + c) + d$
5.	$y = \dfrac{1}{x}$	$y = \dfrac{a}{(bx + c)} + d$

b.   For each of functions 1 through 5, determine if f is symmetric with respect to the y-axis or with respect to the origin. Explain.

2.   Using the functions 3, 4, and 5 above, investigate the effect of adding a constant to the function $y = f(x)$.

a.   Graph the following:

    i. $f(x) = |x|$ ii. $g(x) = |x| + 3$ iii. $g(x) = |x| - 4$

Graph $f(x) = |x|$ on $[-10, 10]$ by $[-6, 6]$[1]. Overlay the graph of $g(x) = |x| + 3$. (What is D?) Overlay the graph $g(x) = |x| - 4$ on the same viewing rectangle ($D = -4$ this time).

b.   Repeat part (2a) for functions 4 and 5, graphing $y = f(x)$, $g(x) = f(x) + 3$, and $g(x) = f(x) - 4$ in each case. Clear the graphics screen before graphing each new set of related functions.

c.   How does the change in D effect the graph of f? Describe the shape and position of the graph of $y = g(x)$ compared to that of $y = f(x)$.

d.   How does the shape and position of the graph of $g(x) = f(x) + D$ compare to that of the graph of $y = f(x)$ when:

    i. $D > 0$ ii. $D < 0$.

e.   Why does the graph of g appear as it does? What effect does D have on the graph of f?

3.   Using functions 1, 3, and 4 from part (1), investigate the effect of changing the parameter A in the function $g(x) = A \cdot f(Bx + C) + D$.

a.   List the values for A, B, C, and D for each of the functions below.

    i. $y = f(x)$ ii. $g(x) = 2 \cdot f(x)$

    iii. $g(x) = 0.25 \cdot f(x)$ iv. $g(x) = -f(x)$

b.   Sketch the functions in part (3a) on the same viewing rectangle. Compare the graphs of these functions.

c.   Describe the shapes and positions of the graphs of the functions in part (3a ii-iv) as compared to the graph of f when:

    i. $|A| > 1$ ii. $|A| < 1$ iii. $A = -1$

d.   What effect does A have on the graph of f? (Consider each of the three cases in part (3c) individually, then discuss their similarities.)

4.   Using functions 3, 4, and 5 from part (1), investigate the effect of changing the parameter C in the function $g(x) = A \cdot f(Bx + C) + D$.

a.   For each of the functions 3 and 5 in part (1), compare the graphs of the following functions. Using an appropriate domain and range, sketch all three functions on the same viewing rectangle.

    i. $y = f(x)$ ii. $g(x) = f(x + 2)$ iii. $g(x) = f(x - 4)$

---

[1] Choose a y-interval that is appropriate for the graphics tool that you are using. If the screen is rectangular with the x-dimension wider than the y-dimension, the y-interval given here might be appropriate. If the screen is square, use a y-interval of $[-10, 10]$. Use the program FAMILY to complete this exploration, using $[-9.4, 9.4]$ by $[-6.2, 6.2]$ (CASIO) or $[-9.6, 9.4]$ by $[-6.4, 6.2]$ (TI).

b. Using function 4, graph $y = f(x + \pi)$, $y = f(x + \frac{\pi}{2})$, $y = f(x - \frac{\pi}{2})$ together on the viewing rectangle [–6.3, 6.3] by [–1.5, 1.5]. Which familiar trigonometric functions (if any) do these graphs resemble?

c. Considering the results from observations made in parts (a) and (b), describe the shape and position of the graph of $g(x) = f(x + C)$ when:

    i. $C > 0$ ii. $C < 0$.

d. Why does the graph of g appear as it does as compared to the graph of f? Describe the effect of first adding C to x then performing the function on this quantity.

5. a. Using functions 3 and 4 from part (1), graph the following on the same viewing rectangle:

    i. $y = f(x)$ ii. $g(x) = f(2x)$
    iii. $g(x) = f(0.25x)$ iv. $g(x) = f(-x)$

b. Which parameter in $g(x) = A \cdot f(Bx + C) + D$ is being investigated this time? Determine the values of A, B, C, and D in $g(x) = A \cdot f(Bx + C) + D$ for each of the functions in part (5a).

c. Compare the shapes and positions of the graphs of the functions in part (5a ii-iv) to the graph of f when:

    i. $|B| > 1$ ii. $|B| < 1$ iii. $B = -1$

d. Why does the graph of g appear as it does, compared to the graph of f? Describe the effect of first multiplying x by B then performing the function on this quantity.

6. The figure to the right shows the graph of a function $y = f(x)$ on the viewing rectangle [–3, 3] by [–3, 3].

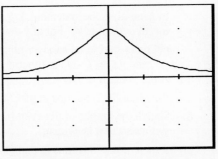

a. Sketch the graphs of each of the following on separate (labeled) axes.

    i. $g_1(x) = f(x + 1)$
    ii. $g_2(x) = .5 f(x + 1)$
    iii. $g_3(x) = .5 f(x + 1) - 3$

b. Is the graph of g_1 helpful in sketching the graph of g_2? Is the graph of g_2 helpful in sketching the graph of g_3? Explain.

c. Sketch the graph of $g_4(x) = -f(x)$. Explain how this graph was determined.

d. Sketch the graph of $g_5(x) = -2 f(x)$. Explain how this graph was determined. How is the graph of g_5 related to the graph of f?

e. Sketch the graph of $g_6(x) = f(-x)$. Explain how this graph was determined. How is the graph of g_6 related to the graph of f?

7. a. Sketch the graphs of the following pairs of functions:

    i. $g(x) = \sin x + \pi/2$ ii. $g(x) = \sin x - \pi$
        $h(x) = \sin (x + \pi/2)$ $h(x) = \sin(x - \pi)$

b. Does $\sin x + k = \sin(x + k)$ for $k = \dfrac{\pi}{2}$? For $k = -\pi$? For any constant k?

Explain using the graphs above as part of your explanation.

c. Sketch the graphs of the following pairs of functions:

    i. $g(x) = 2 \cdot \sin x$ ii. $g(x) = -\sin x$
       $h(x) = \sin(2x)$ $h(x) = \sin(-x)$

d. Does $k \cdot \sin x = \sin(kx)$ for $k = 2$? For $k = -1$? For any constant k? Explain.

8. a. Using functions 2, 3, and 5, compare the graphs of the following pairs of functions:

    i. $g(x) = f(x) + 2$ ii. $g(x) = f(x) - 1$
       $h(x) = f(x + 2)$ $h(x) = f(x - 1)$

  b. For the functions above, does $f(x) + k = f(x + k)$ for $k = 2$? For $k = -1$? For all constants k? For any constant k? Explain using the graphs above as part of your explanation.

  c. In general, is the statement $f(x) + k = f(x + k)$ true for an arbitrary function f for all constants k? For any constant k? Explain.

  d. Using functions 2, 3, and 5, compare the graphs of the following pairs of functions:

    i. $g(x) = 2 \cdot f(x)$ ii. $g(x) = -f(x)$
       $h(x) = f(2x)$ $h(x) = f(-x)$

  e. For the functions above, does $k \cdot f(x) = f(kx)$ for $k = 2$? For $k = -1$? For all constants k? For any constant k? Explain.

  f. In general, is the statement $k \cdot f(x) = f(kx)$ true for an arbitrary function f for all constants k? For any constant k? Explain.

9. A function is said to be an *even function* if, whenever x is in the domain of f, so is $-x$, and $f(x) = f(-x)$. A function is said to be an *odd function* if, whenever x is in the domain of f, so is $-x$, and $f(-x) = -f(x)$.

  a. Which of the functions in part (1) are even?

  b. Sketch the graphs of the even functions in part (1). Describe any symmetry that is evident.

  c. Which of the functions in part (1) are odd?

  d. Sketch the graphs of the odd functions in part (1). Describe any symmetry that is evident.

10. a. The function f provided in the figure at the top of the next page is sketched on the viewing rectangle [-3.2, 3.2] by [-3, 3]. Sketch the graph of $g(x) = f(2x) + 1$.

  b. Does the function $y = f(x)$ in part (10a) appear to be an even function, an odd function, or neither?

  c. For the function $y = f(x)$ given in part (10a), discuss the symmetry of each of the following functions:

    i. $g_1(x) = f(x) + D, D \neq 0$ ii. $g_2(x) = A f(x), A \neq 1$
   iii. $g_3(x) = f(x + C), C \neq 0$ iv. $g_4(x) = f(Bx), B \neq 1$
    Are $y = f(x)$ and $y = g_i(x)$ both even, odd, or neither? Explain.

d. Suppose two or more of the following are false: A = 1, B = 1, C = 0, D = 0. For the function $y = f(x)$ sketched in part (10a), will the function $g(x) = A \cdot f(Bx + C) + D$ be even, odd, neither, or is it necessary to know the specific values of A, B, C, and D before it is possible to determine if g is even, odd, or neither? Explain.

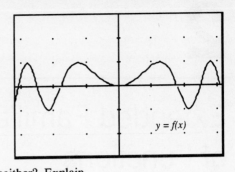

11. a. The function f in the figure to the right is sketched on the viewing rectangle [–3.2, 3.2] by [–1.5, 1.5]. Sketch the graph of $g(x) = -f(x) + \dfrac{1}{2}$.

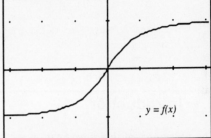

b. Does $y = f(x)$ appear to be an even function, an odd function, or neither? Explain.

c. For the function $y = f(x)$ given in part (11a), discuss the symmetry of each of the following functions:

    i. $g_1(x) = f(x) + D, D \neq 0$ ii. $g_2(x) = A\,f(x), A \neq 1$

   iii. $g_3(x) = f(x + C), C \neq 0$ iv. $g_4(x) = f(Bx), B \neq 1$

Are $y = f(x)$ and $y = g_i(x)$ both even, odd, or neither? Explain.

d. Suppose two or more of the following are false: A = 1, B = 1, C = 0, D = 0. For the function $y = f(x)$ sketched in part (11a), will the function $g(x) = A \cdot f(Bx + C) + D$ be even, odd, neither, or is it necessary to know the specific values of A, B, C, and D before it is possible to determine if g is even, odd, or neither? Explain.

12. a. Let $y = f(x)$ be any even function. Discuss the symmetry of f.

b. For each of (i) through (iv) below, determine whether the following statement is true or false: For an even function $y = f(x)$, the function $y = g_i(x)$ (i = 1, 2, 3, 4) is also an even function for any choice of the parameter A, B, C, or D.

    i. $g_1(x) = f(x) + D, D \neq 0$ ii. $g_2(x) = A\,f(x), A \neq 1$

   iii. $g_3(x) = f(x + C), C \neq 0$ iv. $g_4(x) = f(Bx), B \neq 1$

Explain. Support your conclusion(s). If the statement is true, explain why it is true for all choices of the parameter in question. If the statement is false, give an example of a function for which the statement is false.

c. Let $y = f(x)$ be any odd function. Discuss the symmetry of the function f.

d. Repeat part (12b) for any odd function $y = f(x)$.

2

Extended Families
of Functions

2.1 Introduction

The objectives of this investigation are to determine the relationships between the functions of $y = f(x)$, $y = g(x)$, and

1. $h(x) = f(x) \,\Box\, g(x)$ where $\Box$ is replaced with one of the operations $+$, $-$, $\times$, or $\div$; and
2. $h(x) = f(g(x))$.

How does the new function h relate to the original functions f and g? In particular, how are the domain, range, and symmetries of a function h related to the functions f and g? How is the shape of the graph of h related to the graphs of f and g? Most importantly, what effect does each of the functions f and g have on a function h?

This investigation also includes a review of polynomial and root functions.

2.2 $h(x) = f(x) \,\Box\, g(x)$

How do the functions $f + g, f - g, f \times g$, and $f \div g$ relate to the "parent" functions $y = f(x)$ and $y = g(x)$? Complete the following to investigate these relationships:

1. a. Graph the functions $f(x) = \sin x$, $g(x) = x$, and $h(x) = x + \sin x$ on the viewing rectangle $[-10, 10]$ by $[-10, 10]$, then choose a more appropriate viewing rectangle.

    b.   Compare the vertical distance between $g(x) = x$ and $h(x) = x + \sin x$ with the vertical distance between the x-axis and $f(x) = \sin x$ for several values of x.

    c.   How can the graph of $h(x) = f(x) + g(x)$ be obtained from the graphs of f and g?

2.   a.   Graph the functions $f(x) = \sin x$, $g(x) = x$, and $h(x) = x - \sin x$ using an appropriate viewing rectangle.

    b.   Compare the vertical distance between $g(x) = x$ and $h(x) = x - \sin x$ with the vertical distance between the x-axis and $f(x) = \sin x$ for several values of x.

    c.   How can the graph of $h(x) = f(x) - g(x)$ be obtained from the graphs of f and g?

3.   a.   Graph the functions $f(x) = \sin x$ and $h(x) = x \cdot \sin x$ using an appropriate viewing rectangle.

    b.   Compare the number of oscillations of $f(x) = \sin x$ with those of $h(x) = x \cdot \sin x$ on this viewing rectangle.

    c.   Redraw $h(x) = x \cdot \sin x$. Overlay functions $y = x$ and $y = -x$. What are the largest (smallest) values that $h(x) = x \cdot \sin x$ can achieve for any choice of x? Explain.

4.   a.   Graph functions $f(x) = \sin x$ and $h(x) = \dfrac{1}{\sin x}$ on the viewing rectangle $[-10, 10]$ by $[-6, 6]$.

    b.   Recall that $\dfrac{1}{\sin x} = \csc x$. How are the graphs of $y = \sin x$ and $y = \csc x$ related?

    c.   Why are the asymptotes being formed where they are? Explain.

    d.   Why do these graphs touch at certain points? Explain.

5.   a.   Graph function $h(x) = \dfrac{\sin x}{x}$.

    b.   What is $h(0)$? Explain.

    c.   From the graph, what does $h(0)$ appear to be?

    d.   Zoom in on $h(x) = \dfrac{\sin x}{x}$ a few times to determine the values of $y = h(x)$ for x near 0. What are the x and y end points of this viewing rectangle?

    e.   Describe the appearance of h on this viewing rectangle. What could account for this behavior? Explain in terms of the functions $f(x) = \sin x$ and $g(x) = x$ for x near 0.

6.   From the investigations in parts (1) through (5), how might you determine the graph of $h(x) = f(x) \ \square \ g(x)$ from the graphs of f and g when $\square$ is replaced by:

    a. $+$ or $-$ b. $\times$ c. $\div$

2.3 $h(x) = g(x) \cdot \sin x$

Investigate the effect of multiplying $f(x) = \sin x$ by a constant A, versus multiplying f by a non-constant function $y = g(x)$, as in the following.

1. Consider the effect on the function $f(x) = \sin x$ when f is multiplied by a constant A.

    a. Graph the functions $h(x) = A \sin x$ for A = 2, 0.5, and –3 on a viewing rectangle [–5, 5] by [–3, 3].[1]

    b. How do the graphs of these functions compare with each other?

    c. Overlay $f(x) = \sin x$. How do the oscillations of $f(x) = \sin x$ compare with those of $h(x) = A \cdot \sin x$?

2. Consider the effect on the function $f(x) = \sin x$ when f is multiplied by the non-constant function $g(x) = x^2$.

    a. On a viewing rectangle [–10, 10] by [–100, 100], graph $g(x) = x^2$, $y = -x^2$, and $h(x) = x^2 \cdot \sin x$.

    b. How do the graphs of these functions compare?

    c. Overlay $y = 50 \cdot \sin x$. How do the oscillations of $f(x) = \sin x$ compare with those of $h(x) = x^2 \cdot \sin x$?

3. Consider the effect on the function $f(x) = \sin x$ when f is multiplied by the non-constant function $g(x) = \sqrt{x}$.

    a. On a viewing rectangle [0, 25] by [–5, 5], graph $g(x) = \sqrt{x}$, $y = -\sqrt{x}$, and $h(x) = \sqrt{x} \sin x$.

    b. How do the graphs of these functions compare?

    c. Overlay $f(x) = \sin x$. How do the graphs of these functions compare?

4. From the functions in parts (1) through (3), how are $y = A \cdot \sin x$ and $y = g(x) \cdot \sin x$ similar? How are they different?

2.4 $h(x) = \sin g(x)$

Compare the effect on the graph of $f(x) = \sin (x)$ when the argument of $y = \sin x$ is replaced by a linear function $g(x) = Bx$ versus when the argument $g(x)$ is non-linear, as in the following.

1. Consider the effect on the graph of $f(x) = \sin (x)$ when the argument of $f(x) = \sin x$ is replaced by a linear function $g(x) = Bx$.

---

[1] For the CASIO graphics calculator, use the default viewing window [-4.7, 4.7] by [-3.1, 3.1]. For the TI-81, use [-4.8, 4.7] by [-3.2, 3.1].

    a.   On a viewing rectangle [–5, 5] by [–3, 3], graph $y = \sin 2x$, $y = \sin 5x$, and $y = \sin -3x$.

    b.   Describe the oscillations of $h(x) = \sin (Bx)$ for B = 2, 5, and –3 on this viewing rectangle.

    c.   Compare the graph of $h(x) = \sin (Bx)$ with the graph of $f(x) = \sin x$.

2.   Consider the effect on the graph of $f(x) = \sin x$ when the argument of $f(x) = \sin x$ is replaced by the non-linear function $g(x) = x^2$.

    a.   On a viewing rectangle [–5, 5] by [–1.2, 1.2], graph $h(x) = \sin (x^2)$. Overlay $f(x) = \sin x$.

    b.   Describe the oscillations of $h(x) = \sin (x^2)$ on this viewing rectangle.

    c.   Compare the graph of $h(x) = \sin (x^2)$ with the graph of $f(x) = \sin x$.

    d.   Graph $h(x) = \sin (x^2)$ on [–10, 10] by [–1.2, 1.2]. Why does this function behave the way it does as |x| increases in magnitude?

3.   Replace the argument of $f(x) = \sin x$ with the non-linear function $g(x) = \sqrt{x}$.

    a.   Graph $h(x) = \sin \sqrt{x}$ on a viewing rectangle [0, 100] by [–1.2, 1.2].

    b.   How often does $h(x) = \sin \sqrt{x}$ oscillate on this viewing rectangle?

    c.   Overlay $f(x) = \sin x$. How often does $f(x) = \sin x$ oscillate on this viewing rectangle?

    d.   How does $h(x) = \sin \sqrt{x}$ behave as x increases? Why?

4.   Consider the effect on the graph of $f(x) = \sin (x)$ when the argument of $f(x) = \sin x$ is replaced by the non-linear function $g(x) = 1/x$:

    a.   On a viewing rectangle [–.1, .1] by [–1.2, 1.2], graph $h(x) = \sin (1/x)$.

    b.   Describe the behavior of $h(x) = \sin (1/x)$ on this viewing rectangle.

    c.   Why does this function behave as it does near $x = 0$?

    d.   Change the domain to [–.01, .01] and graph $h(x) = \sin (1/x)$ again. The precision of most graphics tools does not allow a very good picture of this function on this viewing rectangle. How do you think the graph of $h(x) = \sin (1/x)$ should appear? Explain.

    e.   Change the domain to [–3.14, 3.14] and graph $h(x) = \sin (1/x)$ once more. Again, the behavior of the function around $x = 0$ is not shown well. If more points could be plotted, how would it appear? Overlay the graph of $f(x) = \sin x$.

    f.   Why does $h(x) = \sin (1/x)$ behave as it does near $x = 0$?

5.   Considering the work completed in parts (1) through (4), how are the graphs of $h(x) = \sin (g(x))$ for $g(x) = Bx$ versus for non-linear $g(x)$ similar? How are they different?

2.5 Polynomials and Roots

The following questions pertain to the functions given below.

function	$y = f(x)$	function	$y = f(x)$
1.	$x - 1$	3.	$(x - 1)(x + 1)(x + 4)$
2.	$(x - 1)(x + 1)$	4.	$(x - 1)^2(x + 1)(x + 4)$

1. Complete the following for each of the functions 1 through 4.
    a. Graph each of the functions 1 through 4.
    b. Determine the x-intercepts for each function.
    c. Describe the graph of $y = f(x)$ as the function approaches each x-intercept.
    d. Why does the function behave the way it does as it approaches each x-intercept? Use the graph and the algebraic representation of the function in your explanation.

2. a. Graph the opposite of $f(x) = (x - 1)^2(x + 1)(x + 4)$.
    b. What are the x-intercepts?
    c. How is $y = f(x)$ related to $y = -f(x)$? Explain.

3. a. Graph $f(x) = (x - 1)(x + 1)(x + 4)$.
    b. Graph a function that is 2 times f.
    c. Discuss the shape and steepness of these functions as compared to each other.

4. a. Write an algebraic expression for a polynomial function that crosses the x-axis at 3, 2, and –5.
    b. Graph the polynomial function.
    c. Are there other polynomials that satisfy the above conditions? Explain.

5. Write an algebraic expression for a polynomial function that touches (but does not cross) the x-axis at $x = -1$, and crosses the x-axis at 3 and –0.5. Graph this polynomial. Explain your results.

6. a. What are all the possible shapes of a function $y = f(x)$ if it is the product of 2, 3, then 4 linear factors? Explain.
    b. If f is the product of 5 linear factors, what are all of the possible shapes f can have? Sketch these. How can you be sure you have all of the possible shapes?

7. a. Sketch $y = x^3$, $y = x^5$, and $y = x^7$ on [0, 2] by [0, 2].
    b. Discuss the relative magnitudes of the y-values for each of these functions for x in [0, 1] and in [1, 2]. (For more detail, graph these functions on the viewing rectangle [0, 1] by [0, 1] and also on [1, 2] by [1, 10].)
    c. For $x \in [-1, 0]$, which is greater, x^3 or x^5? Explain using the information determined above and information about the symmetry of these functions. (Consider several cases.)

8. a. Sketch $y = \sqrt{x}$, $y = \sqrt[3]{x}$, $y = \sqrt[4]{x}$, and $y = \sqrt[5]{x}$ on [0, 1] by [0, 1] then on [0, 5] by [0, 2].

   b. Discuss the behavior of $y = \sqrt[n]{x}$ on these intervals from what you observed above (Several cases must be considered).

   c. For $x \geq 0$, which is larger, $\sqrt[3]{x}$ or $\sqrt[4]{x}$? Explain.

2.6 Related Problems

1. Graph the following functions. By describing the behaviors of the "parent" functions, explain why these graphs appear as they do on the intervals given.

		x-interval	y-interval
a.	$f(x) = \cos x + x$	[–10, 10]	[–10, 10]
	Also graph $y = \cos x$ and $y = x$.		
b.	$f(x) = x \cos x$	[–6.3, 6.3]	[–7, 7]
	Also graph $y = x$ and $y = -x$.		
c.	$f(x) = \sec x$	[–6.3, 6.3]	[–10, 10]
	Also graph $y = \cos x$.		
d.	$f(x) = x^2 \sin \dfrac{1}{x}$	[–0.1, 0.1]	[–.01, .01]

2. a. Write the polynomial function $p(x) = x^2 + x - 2$ in its factored form. Determine the roots of p and describe its graph before sketching the function.

   b. Sketch the graph of $p(x) = x^2 + x - 2$. Label the axes, indicating the scales for x and y.

   c. From the graph of p, determine the domain of $f(x) = \sqrt{x^2 + x - 2}$.

   d. Sketch $y = f(x)$ on the same axes as $y = p(x)$. Explain why your graph looks as it does by describing the behaviors of $y = p(x)$ and $y = \sqrt{x}$.

3. The graphs of functions f and g are provided in the figure at right.

    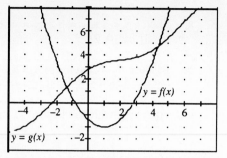

    a. Sketch the graph of
        $$h(x) = \frac{f(x)}{g(x)}$$ on the same set of axes. You may assume that f and g continue to grow as they appear below for $|x|$ large.

    b. Describe the appearance of the graph of $y = h(x)$ for the values of x stated below. For each case, explain why h appears as it does in terms of f and g and the values of the quotients of these functions for x in the intervals given.

        i.   Near $x = a$ (where g crosses the x-axis for $x < 0$)
        ii.   Near $x = b$ (where $f(b) = g(b)$, $b < 0$)
        iii. Near $x = c$ (where f crosses the x-axis, $c < 0$)
        iv.   Near $x = d$ (where f crosses the x-axis, $d > 0$)
        v.   Near $x = e$ (where $f(e) = g(e)$, $e > 0$)
        vi.   As x grows large positively
        vii. As x grows large negatively

    c. On separate (labeled) axes, sketch the graphs of
        i. $h(x) = f(x) + g(x)$
        ii. $h(x) = f(x) - g(x)$

4. Complete the following for each function $y = g(x)$ graphed below.

    a. Sketch the graph of $f(x) = g(x) \cdot \sin x$.

    b. Sketch the graph of $h(x) = \sin(g(x))$.

    c. For each case, explain why the graph appears as it does.

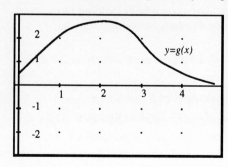

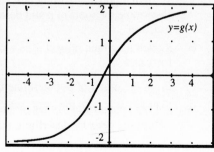

3

Understanding Function Notation

3.1 Introduction

In Chapters 1 and 2, related properties of families and extended families of functions were explored. This exercise provides more work with function notation. It is based on the previous two explorations and requires consideration of the relationship between function graphs and function notation. In particular, the answer to the question, "Is $f(x) + f(h)$ equal to $f(x + h)$?" will be investigated.

3.2 Function Notation

Let f and g be functions defined for all real numbers.

1. Suppose $f(3) = 7$. What point must lie on the graph of f?
2. Suppose the point $(-1, 2)$ lies on the graph of f. What can be inferred about f?
3. Suppose the point $(4, g(4))$ lies on the graph of f. What can be inferred about the relationship between f and g?
4. Suppose the graphs of f and g have a point in common, say (x_0, y_0). What can be inferred about the relationship between $f(x_0)$ and $g(x_0)$?
5. Assuming $f(2) = 3$ and $g(2) = -1$, determine $f(2) + g(2)$. Can $(f + g)(2)$ also be determined? Why or why not?

3.3 Is $f(x + h)$ equal to $f(x) + f(h)$?

1. All of the following problems relate to the function $y = f(x)$ in the figure provided.

    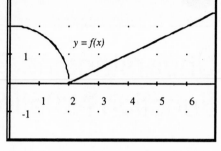

    a. On the axes provided, sketch $g(x) = f(x + .5)$. Label $y = g(x)$.

    b. Locate and label the point $(.5, f(.5))$ on the axes provided.

    c. What is the approximate value of $f(.5)$?

    d. Graph $s(x) = f(x) + f(.5)$, using another color, on the axes above. Label $y = s(x)$.

    e. Is $f(x + h)$ equal to $f(x) + f(h)$? Explain.

2. Suppose $f(x) = \cos(x)$.

    a. Find an expression for $g(x) = f\left(x + \dfrac{\pi}{4}\right)$.

    b. Find an expression for $s(x) = f(x) + f\left(\dfrac{\pi}{4}\right)$.

    c. Sketch the graphs of $y = f(x)$, $y = g(x)$, and $y = s(x)$ all on the same axes using a different color for each. Label each graph.

    d. Is $f(x + h)$ equal to $f(x) + f(h)$? Explain.

3. All of the following problems relate to the function $y = f(x)$ sketched in the figure provided.

    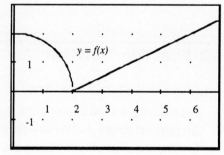

    a. Let h be a small positive number, say in the interval $(0, 0.25)$. Label h on the axes.

    b. On the axes above sketch $g(x) = f(x + h)$.

    c. Locate and label the point $(h, f(h))$ on the axes.

    d. What is the approximate value of $f(h)$?

    e. Graph $s(x) = f(x) + f(h)$, using another color, on the axes above. Label $y = s(x)$.

    f. Is $f(x + h)$ equal to $f(x) + f(h)$? Explain.

4. a. Explain how the information found in each of parts (1e), (2d), and (3f) is related.

    b. How is it different?

    c. Should the answers to these questions be the same? Why or why not?

4

Behavior at a Point
and End Behavior

4.1 Introduction

How does a function behave on an interval over which it is undefined at a single point? Do certain unfamiliar functions behave similarly to more familiar functions at values where they are undefined and as values of $|x|$ grow large? In this exploration, behaviors of functions near a point $x = a$, and as x grows large in absolute value, are investigated both numerically and graphically.

4.2 Behavior at $x = a$ and as $|x|$ Grows Large

1. Graph $f(x) = \dfrac{x^2 - 3x - 4}{x^2 - 1}$ on [-2, 2] by [-10, 10].[1] Note that $y = f(x)$ is undefined at $x = 1$ and $x = -1$.

    a. From the graph, how does f appear to behave near $x = 1$? near $x = -1$?

    b. For values of x (1 or -1) where the graph appears to be continuous (has no visible holes), use the trace option of the graphics utility to estimate the value of y for this choice of x.

    c. To determine the behavior of $f(x) = \dfrac{x^2 - 3x - 4}{x^2 - 1}$ for x near 1 numerically, complete the following table of values. Notice that for each

---

[1] To save time later, enter the code to graph $y = f(x)$ into program 1 (CASIO 7000), f_1 (CASIO 7700) or Y_1 (TI-81).

successive table value, the value of H is half that of the previous entry for H.[2]

H	A + H	f(A + H)	A − H	f(A − H)
0.5	1.5	−5	0.5	7
0.25	1.25	−11	0.75	13
0.125			0.875	
0.0625			0.9375	
0.03125				
0.015625				
0.0078125				
0.00390625				
0.001953125				

d. Observe the magnitude of successive values of y in the table above. Describe the behavior of $f(x) = \dfrac{x^2 - 3x - 4}{x^2 - 1}$ as x approaches 1.

e. Compare the values in the table to the graph of f. How are these related? Explain.

f. To determine the behavior of $y = f(x)$ for values of x near −1, complete the table below.

H	A + H	f(A + H)	A − H	f(A − H)
0.5	−0.5	3	−1.5	2.2
0.25	−0.75	2.714285714	−1.25	2.333333333
0.125			−1.125	
0.0625			−1.0625	
0.03125				
0.015625				
0.0078125				
0.00390625				
0.001953125				

g. Compare the successive y-values listed in part (1f) to the graph of f. Explain how these are related.

h. To obtain more information about the function f, let $f(x) = \dfrac{g(x)}{h(x)}$ where $g(x) = x^2 - 3x - 4$ and $h(x) = x^2 - 1$.

   i. Overlay the graphs of g and h for x in [−5, 5]. How do the graphs of g and h appear on a small interval containing $x = 1$? $x = -1$?

---

[2] Program BEHAVIOR is provided in the appropriate appendix for CASIO and TI calculators. To complete the first two columns, let H = .5. For the last two columns, let H = −.5. In part (1c), let A = 1. In part (1f), let A = −1.

ii. Zoom in on the intersection of the graphs of g and h until both of the graphs appear as straight lines.[3] Trace along the graphs of g and h. What is the ratio of g to h for the value of x where g and h intersect? What does the ratio of g to h appear to be for values of x near the intersection point?

iii. Continue the table below for A = 1 and for A = −1 for H at least 2^{-7}.[4] Compare the values of g and h. Does it make sense that f approaches the value it does as x approaches −1? As x approaches 1? Explain.

H	x=A+H	g(x)	h(x)	f(x)	x=A−H	g(x)	h(x)	f(x)
.5								
.25								
.125								
.0625								

i. The graph of f has a hole at x = −1 and a vertical asymptote at x = 1. From the symbolic form of this function and the work completed in parts (1a) through (1h), can you suggest why the graph behaves as it does? For any rational function, say y = r(x), is it possible to determine for which values of x the graph of r will have an asymptote and for which values of x the graph of r will have a hole? Explain. Suggest a rule for all rational functions.

j. i. Graph $f(x) = \dfrac{x^2 - 3x - 4}{x^2 - 1}$ on [−2, 2] by [−10, 10] again.

ii. Graph f on each of the following domains: [−20, 20], [−200, 200], and [−2000, 2000]

iii. Describe the progression of graphs that were displayed.

iv. What new information is given with each successive graph?

v. For large values of |x|, does the graph of f resemble the graph of another function? What is the similar function? Graph this similar function over the graph of f.

vi. How does y = f(x) behave as |x| grows large?

k. Continue the following table to investigate the behavior of y = f(x) as |x| grows large.[5] The table should go out to at least ±25,000.

x	f(x)	− x	f(−x)
100	0.9696969697	−100	1.02970297
200	0.9849246231	−200	1.014925373
400		−400	

---

[3] It is helpful to choose the ZOOM viewing rectangle so that opposite corners of the rectangle lie on the graph furthest away from the x-axis.

[4] Modify program BEHAVIOR to print out values of more than one function (see appropriate appendix).

[5] In program BEHAVIOR, replace the division sign in the line H÷2→H with a multiplication symbol. Run BEHAVIOR with A = 0, H = ± 100 to complete the table.

l. How does f behave as x grows large? Is this consistent with the graph of f?

m. Graph $g(x) = x^2 - 3x - 4$ and $h(x) = x^2 - 1$ on the domains suggested in part (1j. i-iv) using the range $[0, (Xmax)^2]$. How do the graphs of g and h compare to each other?

n. Add the columns $g(x)$ and $h(x)$ to the table in part (1k), then complete the table. How do the values of g and h compare as $|x|$ grows large?

o. Do the values of f found in part (1k) make sense? Explain.

2. a. Graph $f(x) = \dfrac{x^3 + 7x^2 - x - 25}{x - 3}$ on $[-10, 10]$ by $[-10, 10]$. How does f behave near $x = 3$?

   b. Redraw the graph of f on $[-20, 20]$ by $[-200, 200]$. How does f behave near $x = 3$? Explain why f might behave as it does for x near 3.

   c. Determine the behavior of f as x approaches 3 as in part (1c) above. For this function f, let A = 3 and begin the table with H = 0.5.[2] How does f behave near $x = 3$? Is this behavior consistent with the graph found in part (2b)? Explain.

   d. Repeat part (1h) for $g(x) = x^3 + 7x^2 - x - 25$, $h(x) = x - 3$ and $x = 3$. How do g and h compare with each other on a small interval containing $x = 3$?

   e. Graph f again. Change the range to $[-200, 200]$ by $[-20\,000, 20\,000]$ and graph f once more. For large values of $|x|$, does the graph of f resemble the graph of a familiar function? What is it? Graph this function.

   f. Change the range to $[-2\,000, 2\,000]$ by $[-2\,000\,000, 2\,000\,000]$ and graph f once more. Overlay the function used in part (2e). How does f behave as $|x|$ grows large?

   g. Explain why the asymptote that was visible in part (2b) is not visible on the graphs in parts (2e) and (2f).

   h. Determine the behavior of f numerically as x grows large positively. To do this, create a table as in parts (1k) and (1n) for $y = g(x)$, $h(x)$, and $f(x)$ as x grows large positively. Now add another column to the table showing values of $y = x^2$. What information does this give about $y = f(x)$ for large positive values of x? Is this consistent with the graph of f?

   i. By analyzing the symbolic form of $y = f(x)$, can you suggest why the graph of f behaves like the function $y = x^2$ for large values of $|x|$?

3. a. Graph $f(x) = \sqrt{x^2 - 4}$ on $[-10, 10]$ by $[-1, 10]$. Describe the behavior of $y = f(x)$ near $x = 2$ and $x = -2$.

   b. Determine the behavior of the graph of f as x approaches 2 by completing a table similar to that in part (1c) letting A = 2 and H = 0.5.

   c. Determine the behavior of the graph of f as x approaches -2 by completing a table similar to that in part (1c) letting A = -2 and H = 0.5.

d. How do the numerical values in these tables compare with the graph of
    f near $x = 2$ and $x = -2$?

e. Describe the appearance of the graph of f as x approaches 2 for values of x
    just larger than 2. Also describe the appearance of the graph of f as x
    approaches -2 for values of x just smaller than -2. Why does f behave as
    it does in small intervals of x for $x < -2$ and $x > 2$?

f. Describe the appearance of the graph of f for values of x between -2 and 2.
    Why does the graph of f appear as it does in this interval?

g. Redraw the graph of f on the viewing rectangle $[-100, 100]$ by $[0, 100]$.
    Does this view of f resemble a familiar function? If yes, what is the
    function? Graph this function.

h. Change the range to $[-10{,}000, 10{,}000]$ by $[0, 10{,}000]$. Graph f again.
    Overlay the graph of $y = |x|$. How does the graph of f compare with the
    graph of $y = |x|$ on this viewing rectangle?

i. By analyzing the symbolic form of $y = f(x)$, suggest why f behaves the
    way it does as $|x|$ grows large.

j. Create a table as in part (1k) for $y = f(x)$ as $|x|$ grows large, using
    $x = \pm 100$ and doubling x successively. Is this table consistent with the
    graphs in parts (3g) and (3h)? Explain. Describe the behavior of f as $|x|$
    grows large.

a. Graph $f(x) = \dfrac{\sqrt{x^2 - 3x + 2}}{x + 3}$ on $[-10, 10]$ by $[-10, 10]$. Determine the
    behavior of f near $x = -3$. Clear the graphics screen and sketch
    $g(x) = \sqrt{x^2 - 3x + 2}$ and $h(x) = x + 3$ on the same viewing rectangle.
    Describe the appearance of the graphs of g and h for x near -3.

b. Investigate the behavior of $y = f(x)$ near $x = -3$ by completing a table
    similar to that in parts (1c) and (1h), letting $A = -3$ and $H = 0.5$.
    Describe the behavior of f in terms of the behaviors of g and h as x
    approaches -3.

c. Create a table as in part (1k) for $y = f(x)$ as $|x|$ grows large, using
    $x = \pm 100$ and doubling x successively.

    i.   Describe the behavior of f as $|x|$ grows large.
    ii. Does f seem to be approaching a particular number as x grows large
         positively? What is this number?
    iii. Does f seem to be approaching a particular number as x grows large
         negatively? What is this number?
    iv. Can you create a function that behaves this way as $|x|$ grows large?

d. Sketch g and h on $[-1000, 1000]$ by $[-1000, 1000]$. Describe the
    relationship between the graphs of g and h on this interval. How do these
    graphs relate to the numeric work of part (4c)? Explain.

e. i.   Graph f on $[-100, 100]$ by $[-5, 5]$. Does the graph of f resemble the
         graph of a familiar function? If yes, overlay the graph this function.

    ii. Change the range to $[-1000, 1000]$ by $[-5, 5]$ and graph $y = f(x)$
         again. For large values of $|x|$, does the graph of f resemble the graph

of another function? What is the similar function? Overlay the graph of the similar function.

   iii. By analyzing the symbolic form of $y = f(x)$, suggest why f behaves as it does for $|x|$ large.

5. a. Graph $f(x) = (\sin x)/x$ on the viewing rectangle [–5, 5] by [–3, 3]. Is this function undefined for any choice of x? Explain.

   b. How does $f(x) = (\sin x)/x$ behave near $x = 0$? Trace to the point on the graph where x is as close as possible to 0. Zoom in on the graph several times. What does $f(0)$ appear to be?

   c. Create a table of values for $y = f(x)$ for values of x close to 0, as in part (1c) above. Use A = 0 and H = 0.5.

   d. For a given x, compare the values of $y = f(x)$ and $y = f(-x)$.

   e. Graph $y = \sin x$ and $y = x$ on the viewing rectangle [–1.5, 1.5] by [–1, 1]. As $|x|$ gets very small, how does the graph of $y = x$ compare to that of $y = \sin x$?

   f. Create another table, like that in part (1c), for the function $f(x) = \sin x$, letting A = 0 and H = 0.5.

     i. As x approaches 0, how do the values of x and $\sin x$ compare?

     ii. From the table created in part (5f) and the graphs sketched in part (5e), what would you expect the values of $(\sin x)/x$ and $x/(\sin x)$ to be for x close to zero? Why?

     iii. Does the graph of $f(x) = (\sin x)/x$ make sense, considering your response to part (5f ii)? Explain.

6. a. For all of the functions above, the behavior of the function f was determined at a point $x = a$ where $f(a)$ was either undefined, or was at an endpoint of the function's domain. Summarize work with behavior at a point by suggesting how information from the individual functions g and h helped you understand the behavior of the function $f(x) = \dfrac{g(x)}{h(x)}$ or $f(x) = g(h(x))$.

   b. The behavior of several functions f as $|x|$ grew large was also investigated. Summarize work with behavior at the ends by suggesting how information from the individual functions g and h helped you understand the behavior of the function $f(x) = \dfrac{g(x)}{h(x)}$ or $f(x) = g(h(x))$.

4.3 Related Problems

1. The graphs of functions f and g are provided in the figure on page 21.

   a. Sketch the graph of $h(x) = (g(x)/f(x))$ on the same set of axes. You may assume that f and g continue to grow as they appear in the graph for $|x|$ large.

b. How does $y = h(x)$ behave near $x = a$ (where g crosses the x-axis for $x < 0$)? Explain in terms of f and g and the values of the quotients of these functions for x near a.

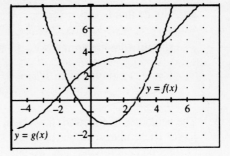

c. How does $y = h(x)$ behave near $x = b$ (where $f(b) = g(b)$, $b < 0$)? Explain.

d. How does $y = h(x)$ behave near $x = c$ (where f crosses the x-axis, $c < 0$)? Explain.

e. How does $y = h(x)$ behave near $x = d$ (where f crosses the x-axis, $d > 0$)? Explain.

f. How does $y = h(x)$ behave near $x = e$ (where $f(e) = g(e)$, $e > 0$)? Explain.

g. How does $y = h(x)$ behave as x grows large positively? Explain.

h. How does $y = h(x)$ behave as x grows large negatively? Explain.

2. a. Sketch the graph of $p(x) = x^2 - 1$. Label the axes, including scales for x and y.

b. From your graph of $y = p(x)$, determine the domain of $f(x) = \sqrt{x^2 - 1}$. Sketch the graph of $y = f(x)$ on the same axes as the graph of p. Explain why your graph looks as it does by describing the behaviors of the individual functions $y = p(x)$ and $y = \sqrt{x}$.

c. Describe the appearance of the graph of f for x near -1 and for x near 1. Explain why f behaves as it does on these intervals.

d. Describe the appearance of the graph of f as x grows large in absolute value.

e. Use the graph of $y = f(x)$ to determine the domain for $h(x) = \dfrac{x - 1}{\sqrt{x^2 - 1}}$.

Explain your choice.

f. Sketch two graphs of $h(x) = \dfrac{x - 1}{\sqrt{x^2 - 1}}$ as follows:

i. The first graph should should indicate any vertical asymptotes or "holes" in the graph and show the behavior of h for both positive and negative values of x for x in the interval $[-6, 6]$. Label the axes indicating the scales for x and y. (Note: You will need to keep the domain of h in mind to get a good graph of $y = h(x)$.) Explain why the graph of h appears as it does on this interval.

ii. The second graph should show the behavior of h as $|x|$ grows large. Label the axes indicating the scales for x and y. Explain why the graph of h appears as it does on this large x-interval.

g. Notice that $h(x) = \dfrac{x-1}{\sqrt{x^2-1}} = \dfrac{x-1}{\sqrt{x-1}\,\sqrt{x+1}}$ can be algebraically

simplified by factoring $\sqrt{x-1}$ from both numerator and denominator. Is

$y = h(x)$ equivalent to $k(x) = \dfrac{\sqrt{x-1}}{\sqrt{x+1}}$ (h with $\sqrt{x-1}$ factored out)?

Explain.

3. The behaviors of functions have been investigated at a point and as $|x|$ grows
large. These investigations naturally lead to the following intuitive definitions
of right-hand, left-hand, and two-sided limits:

Right-Hand Limit:

The notation, $\displaystyle\lim_{x\to c+} f(x) = \mathbf{L}$ means that $f(x)$ can be made arbitrarily

close to L by making x sufficiently close but not necessarily equal to c,
where $f(x)$ is defined for all x in the open interval $(c, c+\delta)$ for $\delta > 0$.

Left-Hand Limit:

The notation, $\displaystyle\lim_{x\to c-} f(x) = \mathbf{L}$ means that $f(x)$ can be made arbitrarily

close to L by making x sufficiently close but not necessarily equal to c,
where $f(x)$ is defined for all x in the open interval $(c-\delta, c)$ for $\delta > 0$.

Two-Sided Limit:

The notation, $\displaystyle\lim_{x\to c} f(x) = \mathbf{L}$ means that $f(x)$ can be made arbitrarily

close to L by making x sufficiently close but not necessarily equal to c,
where $f(x)$ is defined for all x in an open interval containing c, $(c-\delta, c+\delta)$
for $\delta > 0$, but not necessarily at $x = c$.

Use these definitions to answer the following questions.

a. Describe the behavior of $f(x) = x \cdot \cos(1/x^2)$ near $x = 0$, and determine:

i. $\displaystyle\lim_{x\to 0+} x \cdot \cos(1/x^2)$,   ii. $\displaystyle\lim_{x\to 0-} x \cdot \cos(1/x^2)$, and

iii. $\displaystyle\lim_{x\to 0} x \cdot \cos(1/x^2)$.

b. Let $f(x) = (1 - \cos x)/x$. Describe at least two approaches you might use
to investigate whether or not $\displaystyle\lim_{x\to 0} f(x)$ exists.

c. Relate the limit definitions to each of the investigations completed in
Section 4.2 parts (1) through (5), using right-hand, left-hand, and two-sided
limit notation. For example, for part (1), the investigation of the behavior
of $f(x) = \dfrac{x^2 - 3x - 4}{x^2 - 1}$ near $x = -1$, write the limits as follows and

determine the values of these limits:

i. $\displaystyle\lim_{x\to -1+} f(x)$,   ii. $\displaystyle\lim_{x\to -1-} f(x)$, and   iii. $\displaystyle\lim_{x\to -1} f(x)$

5

Continuity

5.1 Introduction

The study of continuous functions is important to the study of calculus. Before the properties of continuous functions can be used to investigate other concepts of calculus, it is important to know what a continuous function is. The following investigation encourages exploration of the relationship between intuitive understanding of continuity and the mathematical definition.

5.2 What is Continuity?

1. What features should a function have to be considered continuous at a point $x = a$? Sketch the graph of a continuous function and describe the features of the graph. Write your own definition of continuity at point $x = a$.

2. Which of the following situations could be classified as examples of continuous quantities? Explain.

   a. Phone bills with respect to time

   b. Amount of change in a person's pocket with respect to time

   c. The amount of syrup in a soda machine over time

   d. Height of the tides in Boston Harbor with respect to time

   e. Change in a person's weight with respect to their height

3. Provide two examples of two related quantities that could be classified as continuous. Explain.

4. The functions listed in the following table have different types of behaviors on an interval containing $x = 0$. For each of the functions, complete the table as shown.

$y = f_i(x)$	Graph of $y = f_i(x)$	$f(0)$	$\lim\limits_{x \to 0} f_i(x)$	Continuous at $x = 0$?
$f_1(x) = x$		0	0	yes, no holes, gaps, or jumps
$f_2(x) = \dfrac{x^2}{x}$		does not exist	0	no, has a hole at the origin, f_2 is not defined there
$f_3(x) = \dfrac{1}{x}$				
$f_4(x) = \dfrac{x}{x}$				
$f_5(x) = |x|$				
$f_6(x) = \dfrac{\sin x}{x}$				
$f_7(x) = \dfrac{|x|}{x}$				
$f_8(x) = \sqrt{x}$				
$f_9(x) = \sin (1/x)$				
$f_{10}(x) = x^2 \cdot \sin (1/x)$				
$f_{11}(x) = \dfrac{1 - \cos x}{x}$				

5. Considering your work in part (4), revise your definition of continuity at a point $x = a$. State your revised definition in terms of limits and function definition at a point $x = a$. Check the definition of continuity in your text. Is your definition consistent with the text definition? Change your definition if necessary.

5.3 Related Problems

1. Consider the function $f(x) = \dfrac{1 - \cos(x)}{x^2}$.

a. Describe at least two ways that you can begin to determine if
$$\lim_{x \to 0} \frac{1 - \cos(x)}{x^2} \text{ exists.}$$

b. Using one of the methods described in part (1a), decide if $\lim_{x \to 0} \dfrac{1 - \cos(x)}{x^2}$
exists or not. If the limit exists, suggest a reasonable value for this limit,
providing reasons for your choice (show all of your work, including
intermediate calculator results). If the limit does not exist, explain why it
does not exist.

c. Is $f(x) = \dfrac{1 - \cos(x)}{x^2}$ continuous at $x = 0$? Explain using the definition of
continuity determined in Section 5.2 part (5).

d. If $f(x) = \dfrac{1 - \cos(x)}{x^2}$ is not continuous at $x = 0$, determine whether or not
the discontinuity is *removable*, in other words, if the discontinuity can be
removed by redefining a single point. If the discontinuity is removable,
determine the value of f at the point of discontinuity that would make the
function continuous. Use the definition of continuity to verify that the
redefined function is now continuous. If this function has some other type
of discontinuity, describe it.

2. a. What is the domain of $h(x) = \sqrt{x^2 - 4}$? Explain your choice.

b. From the domain of $h(x) = \sqrt{x^2 - 4}$, determine the domain of
$$f(x) = \frac{\cos x}{\sqrt{x^2 - 4}} \text{. Explain your choice.}$$

c. Determine $\lim_{x \to 2^+} \dfrac{\cos x}{\sqrt{x^2 - 4}}$.

   i. How did you determine the behavior of this function for x near 2?
   Describe the tool(s)/method(s) you used, and how you used them.
   ii. Explain why $y = f(x)$ behaves as it does for values of x just greater
   than 2. Describe your results in terms of the functions that make up
   $y = f(x)$.

d. For what values of x is $f(x) = \dfrac{\cos x}{\sqrt{x^2 - 4}}$ continuous? Explain your
response in terms of the function in the numerator, and the composite
function in the denominator (you must address each function separately).

e. Describe the behavior of $f(x) = \dfrac{\cos x}{\sqrt{x^2 - 4}}$ as $|x| \to \infty$.

   i. How did you determine the behavior of this function for large $|x|$?
   Describe what tools(s)/method(s) you used, and how you used them.
   ii. Explain why $y = f(x)$ behaves as it does for large values of $|x|$.
f. Determine any vertical asymptotes for f. Explain your work.

6

Understanding
the Derivative

6.1 Introduction

To understand the derivative, we investigate the meaning of rate of change in the familiar setting, distance versus time. The ideas related to rate of change will be modeled graphically. The appearance of graphs of functions will be studied over small intervals for x, in Section 6.3 *Derivatives Locally* and over longer intervals for x, in Section 6.4 *Definition of the Derivative*.

6.2 Rate of Change—Distance Versus Time

Commuters are often interested in how fast they must travel to get from one place to another, knowing the amount of time available to complete the journey, and the distance to be traveled. This interest increases if the time is shorter than usual, and several police officers are stationed along the route. This part of the investigation explores the relationship between distance and time on large, then small intervals of time. In particular, the commuter's point of view of velocity will be compared with that of an officer of the law.

COMMUTER'S POINT OF VIEW

1. a.   How would a commuter determine average velocity for the particular commute given below?

	Time	Odometer (in miles)
Initial	2:30 p.m.	62,137.3
Final	3:12 p.m.	62,172.4

b. What is the average velocity of the commuter over the time interval given?

2. Sketch a possible graph of the function $y = f(t)$ for the commuter's travel where initial and final times and distances are given in the table above. Keep in mind that the commuter must stop at traffic lights, stop signs, etc. and cannot travel at a constant speed from the moment travel begins until arrival at a final destination.

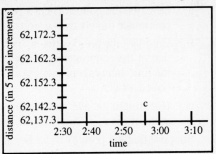

a. Describe the commuter's velocity according to the graph you've drawn. Your description should include reasons for the commuter's change in velocity at various time intervals.

   i. Are there certain time intervals during which the commuter seems to not be traveling any distance at all? Label these times on your graph. What is the appearance of the graph at such times?

   ii. What might be happening on each of these intervals of time where the graph is not rising?

b. How does the commuter's average velocity, found in part (1), relate to the graph sketched above?

c. On the graph above, draw a line containing the points that indicate the initial and final times and positions. How does the velocity found in part (1) relate to this line?

d. On the graph above, what is the slope of the line containing the points (2:30, 62,137.3) and (3:12, 62,172.4)?

e. How does this value compare with the average velocity found in part (1)?

3. Suppose $t = a$ represents the initial time and $t = b$ the final time over which travel took place.

a. Label these values on the axes of the graph above.

b. Assuming distance is a function f of time (distance changes according to time), what do $f(a)$ and $f(b)$ represent?

c. Over the time interval $[a, b]$, where a position function is given by $y = f(t)$, write an expression for average velocity.

d. Does this quantity look familiar? What is this quantity in terms of the graph of f?

e. The line sketched in part (2c) contains the points $(a, f(a))$ and $(b, f(b))$. If $y = f(t)$ represents distance traveled, how might the average velocity over a given time interval of a distance versus time function be determined if the function is given graphically?

POLICE OFFICER'S POINT OF VIEW

A police officer patrolling a road over which a commuter must travel has a different point of view from that of the commuter.

1. a. Describe the police officer's perception of a commuter's velocity as the commuter travels over a stretch of road that the police officer is patrolling.

    b. Compare the police officer's perception of the commuter's velocity with that of the commuter's perception of his or her average velocity. Compare the time intervals of interest to the commuter with those of interest to the police officer.

    c. How might the average velocity of the commuter, from the police officer's point of view, be determined? How might this be done accurately?

2. Suppose that the police officer begins observing the commuter at time c (labeled on graph in part (2) above). He observes the commuter's travel for a very short period of time, say h seconds.

    a. What is the final time?

    b. What are the distances corresponding to the initial and final times?

    c. Write an algebraic expression for the average velocity over this short time interval.

3. The graph provided depicts a distance vs. time function. Time c is indicated. For h = 4 seconds, sketch a line containing $(c, f(c))$ and $(c-4, f(c-4))$.

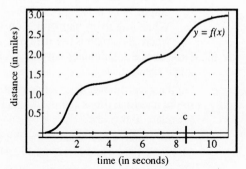

    a. Such lines are called secant lines. Draw secant lines containing $(c, f(c))$ and $(c + h, f(c + h))$ for h = 3, 2, 1, and 0.1.

    b. As h decreases in magnitude, how do the secant lines compare with the curve of $y = f(t)$ on a small interval containing c?

    c. Over the same time intervals as in part (3b), how well would the values of the slopes of successive lines approximate the velocity of the commuter at time $t = c$?

SUMMARY

In the investigation just completed, average velocity was modeled both graphically and symbolically over large and small intervals. Summarize your observations. The following are particularly important.

1. How do the symbols giving average velocity relate to the graph of a distance versus time function $y = f(t)$? Explain.

2. Describe the relationship between secant lines containing two points $(a, f(a))$ and $(b, f(b))$ on the graph of $y = f(t)$, and on the graph of $y = f(t)$ when

    a. la − bl, the distance between t-values a and b, is large, and

    b. la − bl is very small.

3. How might the slopes of secant lines be useful in determining the average velocity of an object over any given time interval?

4. Combine the information from parts (1), (2), and (3) of the *Police Officer's Point of View* to describe how instantaneous velocity might be estimated. (Instantaneous velocity is the velocity of an object at a particular point in time, rather than over an interval of time.)

5. Discuss the possibility of using the limit definition in Chapter 4 to write the instantaneous velocity of the commuter at time $t = c$.

6.3 Derivatives Locally

To obtain information about how a function behaves over very small intervals of x, the graph of f will be magnified several times. This information is referred to as local information about the graph of a function $y = f(x)$. To further explore the relationship between velocity and the graph of a function, we will investigate the graphs of the following functions on very small intervals of x. [1]

function	$y = f(x)$	function	$y = f(x)$
1.	$\sqrt{4 - x^2}$	4.	$|x|$
2.	$\sin \dfrac{1}{x}$	5.	$\left| \dfrac{\cos 25x}{25} \right| + x$
3.	$x^5 - 5x^3$		

VELOCITY GRAPHICALLY

1. Complete the following for $f(x) = \sqrt{4 - x^2}$.

    a. Graph $y = f(x)$ on a viewing rectangle $[-5, 5]$ by $[-3, 3]$. [2]

---

[1] To save time later, enter the graph statement for the function being investigated into Program 1 (CASIO 7000), f_1 (CASIO 7700), or into Y_1 (TI).

[2] Graph the function on the default viewing rectangle (CASIO) or $[-4.8, 4.7]$ by

b. To investigate the behavior of $y = f(x)$ around the point $(1, f(1))$, zoom in several times on the graph of $y = f(x)$ on a small interval containing $x = 1$. Sketch the graph of f when the x-interval shown on the screen is no wider than 0.02. Label the axes. (Note: To minimize the distortion to the slope of the function, try to zoom in on the function using the same zoom factor for both x and y.)

c. If you were to draw a line containing the end points of the function shown on the screen (xmin, f(xmin)) and (xmax, f(xmax)), how well would this line approximate the curve of $y = f(x)$ for small values of (xmax − xmin)?

d. Use the trace option of the graphics tool to determine the values of the following (DO NOT ROUND THESE VALUES!):

x m i n _____ f(x m i n) _____ x m a x _____ f(x m a x) _____

e. Sketch the line segment whose endpoints are the values just found, over the graph of f. How does the line through (xmin, f(xmin)) and (xmax, f(xmax)) compare with the graph of $y = f(x)$ on the last interval found?

f. If $y = f(x)$ is a distance-versus-time function, then average velocity of f over the interval [xmin, xmax] is given by

$$v = \frac{\text{change in distance}}{\text{change in time}} = \frac{f(\text{xmax}) - f(\text{xmin})}{\text{xmax} - \text{xmin}}.$$

i.   How does the value of this expression relate to the line described in part (1c)?

ii. How well is the velocity at $x = 1$ approximated by the slope of the line containing the points (xmin, f(xmin)) and (xmax, f(xmax)) on a very small interval containing $x = 1$?

2. Repeat parts (1a), (1b), (1c), and (1f) for functions 2, 3, 4, and 5.[3]

3. Complete the following for $f(x) = \sin \dfrac{1}{x}$.

a. Graph $f(x) = \sin \dfrac{1}{x}$ on [−.1, .1] by [−1.2, 1.2].

b. Graph $f(x) = \sin \dfrac{1}{x}$ on [−.001, .001] by [−1.2, 1.2]. Describe the graph of f on this interval. Regraph f on [.0001, .001], keeping the range at [−1.2, 1.2]. Sketch the graph labeling the domain and range on the axes. How does f appear on this viewing rectangle? Zoom in on the graph of f several times until the graph appears as a straight line. Sketch the graph, labeling the axes for x and y.

---

[3] [−3.2, 3.1] (TI). If the graphics tool has a square screen, use [−5, 5] by [−5, 5].
To save time later, enter the graph statement for the function being investigated into Program 1 (CASIO 7000), f_1 (CASIO 7700), or into Y_1 (TI).

c. For $f(x) = \sin \dfrac{1}{x}$ on the last interval sketched, how well is the average velocity over the interval [xmin, xmax] approximated by the slope of the line containing the points (xmin, f(xmin)) and (xmax, f(xmax)) on a very small interval for x?

d. Is the graph of f locally straight on any small interval containing $x = 0$? Explain in terms of the behavior of $g(x) = \sin x$ and $h(x) = \dfrac{1}{x}$ for x near 0.

e. If $x = 0$ is not an element of an interval [a, b], is it possible to find a value c in [a, b] for which the graph of f over [a, c] is locally straight? Explain.

4. Complete the following for $f(x) = |x|$.

a. If the function $y = f(x)$ describes the position of an object over time, describe how the object would be moving over an interval containing $x = 0$.

b. What object might move in the way you described in part (4a)?

c. For any small interval containing $x = 0$, how well would an average velocity reading for that interval compare to the actual velocity that the object is traveling at any moment during the interval?

d. Is it possible to determine the velocity that the object is traveling at exactly time $x = 0$?

5. Complete the following for $f(x) = \left| \dfrac{\cos 25x}{25} \right| + x$.

a. Sketch $y = f(x)$ on the interval $[0, \dfrac{\pi}{5}]$ by $[0, \dfrac{\pi}{5}]$. In the interval $[0, \dfrac{\pi}{5}]$, for what values of x does the graph of f change directions suddenly? Explain.

b. Zoom in on the graph of f around a point where the graph changes direction suddenly. Sketch the graph. Label the x and y scales.

c. If the function $y = f(x)$ describes the position of an object over time, describe how the object would be moving over an interval in which the graph changes direction suddenly.

d. Could an object move in the way you described in part (5c)?

e. For any small interval containing a sudden change in direction on the graph, how well would an average velocity reading for that interval compare to the actual velocity that the object is traveling at any moment during the interval?

6. a. In parts (1) and (2), the behavior of the graph of $y = f(x)$ on an interval containing $x = 1$ was investigated for the five functions listed above. Are there other values for $x = a$ for which the graph of f would behave similarly? Explain.

b. For very small x-intervals [a, b] containing $x = c$ over which a function is defined and continuous, how well is the velocity at time $x = c$

approximated by the slope of the line containing the points (a, f(a)) and
(b, f(b))? Explain. (This requires a multiple-part answer.)

6.4 Definition of the Derivative[1]

The behavior of secant lines to the graph of a given function through a point
(c, f(c)) are numerically and graphically investigated.[2]

1. To observe the appearance of secant lines through the points (c, f(c)) and
    (c + h, f(c + h)) on the graph of $f(x) = \sin x$ as h gets small, graph $y = \sin x$
    on the viewing rectangle [−1, 3.7] by [−1.55, 1.55].

    a. For each choice of h in the table below,

        i.   Sketch secant lines through the points (c, f(c)), (c + h, f(c + h)) where
             $c = \dfrac{\pi}{2}$.

        ii. Describe the appearance of successive secants to the graph of
             $f(x) = \sin x$ through the points $(\dfrac{\pi}{2}, 1)$ and $(\dfrac{\pi}{2} + h, \sin(\dfrac{\pi}{2} + h))$ as
             h approaches 0 for h < 0.

        iii.. In the table below, record the slope of the secant line,

             $$m = \frac{f(c + h) - f(c)}{(c + h) - c} = \frac{f(c + h) - f(c)}{h},$$

             for each choice of h. (Recording values of c + h and f(c + h) is
             optional. Columns are included to assist persons completing
             calculations without a program to compute the slope.)

h	c + h	f(c + h)	m
−2	$\dfrac{\pi}{2}$ − 2 = −0.4292037		
−1			
−0.5			
−0.25			
−0.125			
−0.0625			
−0.03125			
−0.015625			

---

[1] A graphics package with a program to graph secant lines and exhibit the values of
their slopes will be helpful in completing this investigation.

[2] For CASIO and TI graphics calculator users, the program SECANT will be used to
complete this investigation. Functions to be investigated must be entered into Y$_1$ (TI)
or f$_1$ (CASIO 7700). For the CASIO 7000, the function to be evaluated must be entered
into Program 0, its graph into Program 1.

iv. What number do the values of the slopes appear to be approaching for h < 0?

b. For each choice of h in the table below,

   i. Sketch secant lines through the points $(c, f(c))$, $(c + h, f(c + h))$ where $c = \dfrac{\pi}{2}$.

   ii. Describe the appearance of successive secants to the graph of $f(x) = \sin x$ through the points $(\dfrac{\pi}{2}, 1)$ and $(\dfrac{\pi}{2} + h, \ \sin(\dfrac{\pi}{2} + h))$ as h approaches 0 for h > 0.

   iii. In the table below, record the slope m of the secant line for each choice of h.

h	c + h	f(c + h)	m
2	$\dfrac{\pi}{2} + 2 = 3.5707963$		
1			
0.5			
0.25			
0.125			
0.0625			
0.03125			
0.015625			

   iv. What number do the values of the slopes appear to be approaching for h > 0?

c. How do the answers to parts (1a iv) and (1b iv) compare? Does this make sense? Explain.

d. In part (1a), by comparing slopes of successive secant lines as h approaches 0 from values of h less than 0, the following limit was investigated:

$$\lim_{h \to 0^-} \frac{f\left(\dfrac{\pi}{2} + h\right) - f\left(\dfrac{\pi}{2}\right)}{h}$$

In part (1b), by comparing slopes of successive secant lines as h approaches 0 from values of h greater than 0, the following limit was investigated:

$$\lim_{h \to 0^+} \frac{f\left(\dfrac{\pi}{2} + h\right) - f\left(\dfrac{\pi}{2}\right)}{h}$$

What do these investigations suggest about

$$\lim_{h \to 0} \frac{f\left(\frac{\pi}{2} + h\right) - f\left(\frac{\pi}{2}\right)}{h} ?$$

e.   On the same axes, sketch the graphs of

    i. $f(x) = \sin x$,

    ii. the secant line containing the points $(c, f(c))$ and $(c + h, f(c + h))$ with $C = \frac{\pi}{2}$ and $H = 0.01$, and

    iii. the secant line containing the points $(c, f(c))$ and $(c + h, f(c + h))$ with $C = \frac{\pi}{2}$ and $H = -0.01$.

f.   In how many points do the graphs of the secant lines through

$(\frac{\pi}{2}, \pm 0.01, \sin\left(\frac{\pi}{2} \pm 0.01\right))$ appear to intersect the graph of

$f(x) = \sin x$? Explain.

g.   A line that intersects a circle in a single point is called a tangent line. How would a secant line to a curve $y = f(x)$ through the points $(c, f(c))$ and $(c + h, f(c + h))$ as h approaches zero compare to a tangent line to f through $(c, f(c))$?

The definition of a tangent line is generalized to any curve (not just circles) by defining its slope as follows.

Definition of the Derivative

If $y = f(x)$ is a function, then the derivative of y with respect to x at $x = c$ is

$$f'(c) = \lim_{h \to 0} \frac{f(c + h) - f(c)}{h}$$

if this limit exists. When the limit does exist, it is called the instantaneous rate of change of y with respect to x at $x = c$. In graphic terms, $f'(c)$ is the slope of the tangent line to the graph of $y = f(x)$ at the point $(c, f(c))$. Note: If we write $y = f(x)$, $f'(c)$ is also called the derivative of y with respect to x at $x = c$.

2.   Graph the function $f(x) = |4 - x^2|$ on the viewing rectangle $[-5, 5]$ by $[-2, 5]$. Explain why the graph of the function appears as it does, especially around $x = -2$ and $x = 2$.

    a.   Regraph $f(x) = |4 - x^2|$ on the viewing rectangle $[-5, 0]$ by $[-1, 12]$.

       i.   Repeat the investigation of part (1a), drawing successive secants between the points $(c, f(c))$ and $(c + h, f(c + h))$ for $c = -2$ and $h = -2$. Record the values of h and the corresponding slopes of the secant lines in a table. When at least five secant lines have been drawn, view the graphics screen.

       ii. What value do the slopes of the secant lines appear to be approaching as h approaches 0 for $h < 0$?

b. i.   Without clearing the graphics screen, repeat part (2a) with c = –2 and h = 2.

   ii.   What value do the slopes of the secant lines appear to be approaching as h approaches 0 for h > 0?

c. Compare the results of parts (a) and (b).

d. In part (2a), the following limit was investigated graphically and numerically:

$$\lim_{h \to 0^-} \frac{f(-2+h) - f(-2)}{h}$$

In part (2b), the following limit was investigated:

$$\lim_{h \to 0^+} \frac{f(-2+h) - f(-2)}{h}$$

What do these investigations suggest about

$$\lim_{h \to 0} \frac{f(-2+h) - f(-2)}{h}$$? Explain.

e. Does the derivative exist at $x = -2$ for $f(x) = |4 - x^2|$? Explain using the definition of derivative.

f. Does $f'(2)$ exist for $f(x) = |4 - x^2|$? Explain using the definition of derivative and both a graphical argument and a numerical repeating parts (2a) and (2b) above with c = 2, h = –2 and then with c = 2, h = 2.

6.5 Related Problems

1. Let f be the function whose graph is provided in the figure at right. Suppose that a is a number and h is a positive number as shown.

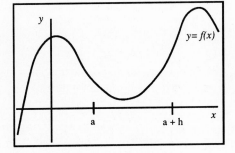

a. Label the points $P(a, f(a))$ and $Q(a + h, f(a + h))$ on the graph of f and draw the secant line between these two points.

b. Write an expression for the slope of the secant line through the points P and Q.

c. Write an expression (involving a limit) for the slope of the line tangent to the graph of f at the point $P(a, f(a))$.

d. Write an equation for the tangent line through the point $(a, f(a))$.

2. In geometry, a tangent is defined as any line that intersects a circle in only one point. The definition of a tangent line in calculus is that line for which the slope of the line through the point (c, f(c)) is given by the expression

$$f'(c) = \lim_{h \to 0} \frac{f(c + h) - f(c)}{h}$$

Is a tangent line as defined in geometry still a tangent line as defined in calculus? Explain.

3. Consider the following piecewise defined function:

$$f(x) = \begin{cases} \sin(\pi x) & \text{for } x \leq 2.5 \\ x^2 - 5x + 7.25 & \text{for } x \geq 2.5 \end{cases}$$

a. Graph the function on the axes provided to the right. Does the derivative of the function $y = f(x)$ appear to exist at $x = 2.5$? Explain.

b. Numerically determine the derivative of $y = f(x)$ at $x = 2.5$ if it exists. Show and explain your work.

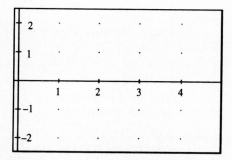

4. A table of values for a certain differentiable function is provided on the right. Use this information to determine an approximation for $f'(1)$. Clearly describe the method you use to determine your approximation.

x	$f(x)$
0.8	0.527
0.9	0.445
1	0.368
1.1	0.298
1.2	0.237

7

Hierarchy of Functions

7.1 Introduction

How are the concepts of differentiation, continuity, and the existence of a limit at a point $x = a$ related? It is possible to categorize functions into a hierarchy, depending on which of these properties each function possesses. This exercise allows you to consider functions that possess the properties of differentiability, continuity, and/or the existence of a limit at a point, and to consider the relationships among these properties.

7.2 Relating Differentiation, Continuity, and Existence of Limits

Complete the following for each of the statements below:

a. Determine whether the statement is true or false.

b. If the statement is true, explain why it is true. If it is false, give a function (graphically or symbolically) for which the statement fails.

c. Explain how the function found for part (2) satisfies the given condition of the statement and fails to satisfy the conclusion.

1. If $\lim\limits_{x \to a} f(x)$ exists, then f is continuous at a.

2. If $\lim\limits_{x \to a} f(x)$ exists, then f is differentiable at a.

3. If f is continuous at a, then $\lim\limits_{x \to a} f(x)$ exists.

4. If f is continuous at a, then f is differentiable at a.

5. If f is differentiable at a, then $\lim\limits_{x \to a} f(x)$ exists.

6. If f is differentiable at a, then f is continuous at a.

7.3 Related Problems

1. The function $f(x) = \dfrac{3x^2 + 2x - 1}{x + 1}$ is defined and continuous for all x except
    $x = -1$. Is it possible to define $f(-1)$ so that f is continuous at $x = -1$? Explain
    your answer.

2. a. Describe the appearance of the
        graph of a continuous function
        that fails to be differentiable at
        $x = 3$. (Note: There are at least
        two possibilities.) Use both
        the definition of continuity and
        the definition of the derivative
        to explain your choice.

    b. Sketch such a graph.

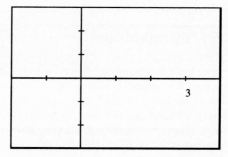

8

Graphical Differentiation

8.1 Introduction

In Chapter 6, we explored the relationship between the rate of change and the graph of a function. In this exploration, the information about slopes of tangent lines to the graph of a function will be used to determine rules for differentiation of certain common functions.

Recall the **Definition of the Derivative:**

If $y = f(x)$ is a function, then **the derivative of y with respect to x at $x = c$ is**

$$f'(c) = \lim_{h \to 0} \frac{f(c + h) - f(c)}{h}$$

if this limit exists. When the limit does exist, it is called the instantaneous rate of change of y with respect to x at $x = c$. In graphic terms, $f'(c)$ is the slope of the tangent line to the graph of $y = f(x)$ at the point $(c, f(c))$. Note: If we write $y = f(x)$, $f'(c)$ is also called the derivative of y with respect to x at $x = c$.

It is possible to estimate $y = f'(x)$ graphically given the graph of $y = f(x)$. Some *landmarks* of f which will be helpful in sketching f' follow.

1. Consider values of x where the slope of the tangent is zero:
   a. Describe characteristics of the graph of a function f on an interval containing $x = c$ for which the slope of the tangent through the point $(c, f(c))$ is zero.
   b. If $f'(c) = 0$, what point is on the graph of $y = f'(x)$?

2. Consider intervals where the slopes of the tangents to $y = f(x)$ are negative:

    a. Describe the appearance of the graph of f on an interval over which $f'(x) < 0$.

    b. If the slope of $y = f(x)$ is negative on an interval, then $f'(x)$ $\boxed{<}$ $\boxed{>}$ $\boxed{=}$ 0 (circle one). If $f'(x) < 0$ on an interval, where does the graph of $y = f'(x)$ lie with respect to the x-axis?

3. Consider intervals where the slopes of the tangents to $y = f(x)$ are positive.

    a. Describe the appearance of the graph of f on an interval over which $f'(x) > 0$.

    b. If the slope of $y = f(x)$ is positive on an interval, then $f'(x)$ $\boxed{<}$ $\boxed{>}$ $\boxed{=}$ 0 (circle one). Where does the graph of $y = f'(x)$ lie with respect to the x-axis for x in such intervals?

8.2 Sketching Derivative Functions by Hand

1. Below are two pairs of axes. Use the axes (2) provided to sketch the graph of the derivative of $y = f(x)$, where f is the function given in graph (1). Sketch the derivative of $y = f(x)$ as follows.

    a. Draw several tangent lines along the curve of f.

    b. Estimate the slope f' of each tangent, using the grid marks as a guide.

    c. Plot the points $(x, f'(x))$ on graph (2).

    d. Connect the points to sketch the graph of $y = f'(x)$.

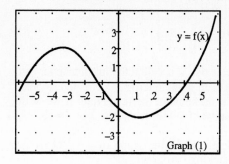

Graph (1)

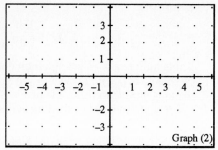

Graph (2)

2. The graphs of several common functions $y = f(x)$ are sketched below. Complete the following for each function.

    a. Sketch the graph of the derivative f' on the same axes as f, following the suggestions given in part (1). Briefly explain why your graph is sensible.

    b. Conjecture a rule to determine the derivative of the given function $y = f(x)$. Provide reasons for your choice.

c. For functions $f(x) = c$ and $f(x) = mx + b$, use the limit definition of the derivative to verify that the rule you have conjectured in part (b) is correct.

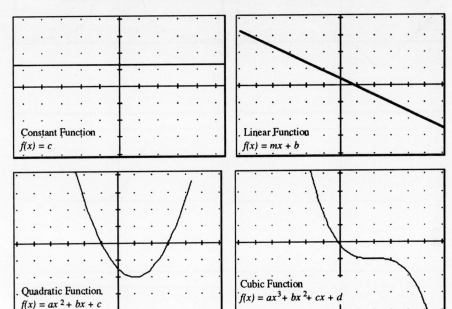

Constant Function
$f(x) = c$

Linear Function
$f(x) = mx + b$

Quadratic Function
$f(x) = ax^2 + bx + c$

Cubic Function
$f(x) = ax^3 + bx^2 + cx + d$

3. Among the functions g_1 through g_{10}, find the graph of the derivative of each function f_1 through f_5. Briefly explain your choice in each case.

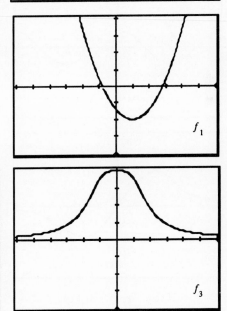

f_1

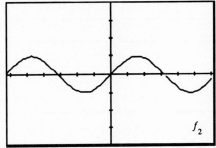

f_2

f_3

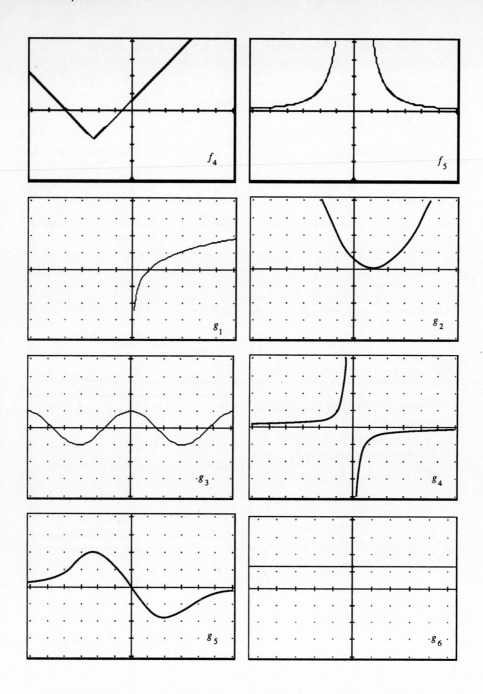

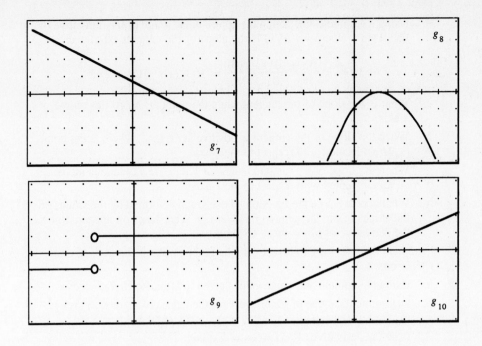

8.3 More Derivative Functions

1. Complete the following to
    determine the function that is the
    derivative of $f(x) = \sin x$.

    a. Graph $f(x) = \sin x$ on the
        viewing rectangle $[-6.3, 6.3]$
        by $[-2, 2]$ and also on the axes
        provided below.[1]

    b. Sketch tangent lines along the
        graph of $y = f(x)$. Estimate the
        slopes of the tangent lines
        drawn. For each tangent drawn along the curve of f through $(x, f(x))$, plot a
        point $(x, M(x))$.

    c. Mark the x-axis in units of $\pi/4 \approx 0.785$. What is the slope of the tangent
        for the x values: $-3\pi/2$, $-\pi$, $-\pi/2$, 0, $\pi/2$, π, and $3\pi/2$? Mark the points
        $(x, M(x))$ on the graph provided.

---

[1] Use program TANGENT (found in the appropriate appendix) to complete this
investigation. Let $A = -2\pi$ and $B = 2\pi$.

d. Using a graphics tool, sketch the approximate graph of the derivative of $f(x) = \sin x$ where $y = M(x)$ is the slope of the tangent estimated by

$$M(x) = \frac{f(x + .01) - f(x)}{.01} = 100(f(x + .01) - f(x)).$$

e. What is the slope of the tangent when the point $(x, M(x))$ lies on the x-axis?

f. When the slope of the tangent is positive at $(x, f(x))$, where is the point $(x, M(x))$ located with respect to the x-axis?

g. When the slope of the tangent is negative at $(x, f(x))$, where is the point $(x, M(x))$ located with respect to the x-axis?

h. What appears to be the function whose graph is formed by the values of the slopes of the tangents to $f(x) = \sin x$? What function appears to be the derivative of $f(x) = \sin x$? Explain.

2. Complete the following to investigate the function that is the derivative of $f(x) = \cos x$.

a. Using the domain and range values given in (1a), sketch $f(x) = \cos x$.

b. Determine the slope of $f(x) = \cos x$ for the values of x given in (1c).

c. Sketch $y = M(x)$ where $f(x) = \cos x$ (as in part (1d)). What function appears to be the derivative of $f(x) = \cos x$? Explain.

3. Complete the following to investigate the function that is the derivative of $f(x) = |x|$.

a. Sketch $f(x) = |x|$ on the viewing rectangle $[-3, 3]$ by $[-3, 3]$.[1]

b. Determine the slope of the lines tangent to the graph of $f(x) = |x|$ for $x = -2, -1, 1, 2$.

c. Does the graph of $f(x) = |x|$ have a tangent line at $x = 0$?

d. Recall the limit definition of $f'(0)$:

$$f'(0) = \lim_{h \to 0} \frac{f(0 + h) - f(0)}{h} = \lim_{h \to 0} \frac{|0 + h| - 0}{h} = \lim_{h \to 0} \frac{|h|}{h}.$$

i. What is $\displaystyle\lim_{h \to 0^+} \frac{|h|}{h}$?

ii. What is $\displaystyle\lim_{h \to 0^-} \frac{|h|}{h}$?

iii. What is $f'(0) = \displaystyle\lim_{h \to 0} \frac{|h|}{h}$?

iv. Does $f'(0)$ exist? Explain.

v. According to the definition of the derivative, is there a tangent line at $x = 0$? Explain.

e. Sketch $y = M(x)$ for $f(x) = |x|$. Explain the appearance of the graph of M around $x = 0$. Is the graphics tool providing an accurate graph? Why or why not?

---

[1] Run program TANGENT with $A = -3$ and $B = 3$.

f.   Write an algebraic expression for the derivative of $f(x) = |x|$. Note: It is permissible to write $y = f'(x)$ as a piecewise function.

8.4 Related Problems

1. Sketched in the figure at right is the graph of a function f. Suppose another function g has the following properties:

$$g(-1) = -2$$

and $g'(x) = f'(x)$

for all x. Sketch the graph of g using the same axes. Briefly explain your results.

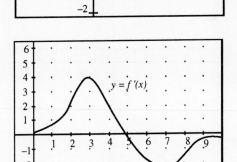

2. The graph of $y = f'(x)$ is shown in the figure at right. Assume that the function $y = f(x)$ represents the number of sales of an innovative calculus workbook in its first several years. **Note that this is the graph of the derivative of f, not the graph of the function f itself.**

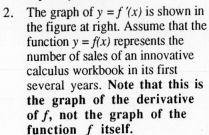
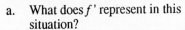

   a.   What does f' represent in this situation?

   b.   Estimate for which year(s) sales were at a maximum. Explain.

   c.   Estimate for which year(s) the least number of workbooks were sold? Explain.

   d.   During which year(s) were the sales increasing most rapidly? Explain.

   e.   If you were the publisher, would you have chosen to carry the workbook for the number of years represented in the graph? Explain.

3. Consider the graphs of functions f_1 through f_6 which follow. For each function graphed

   a.   sketch the graph of its derivative,

   b.   sketch the graph of a function for which this function is the derivative, and

   c.   explain briefly how each graph was determined.

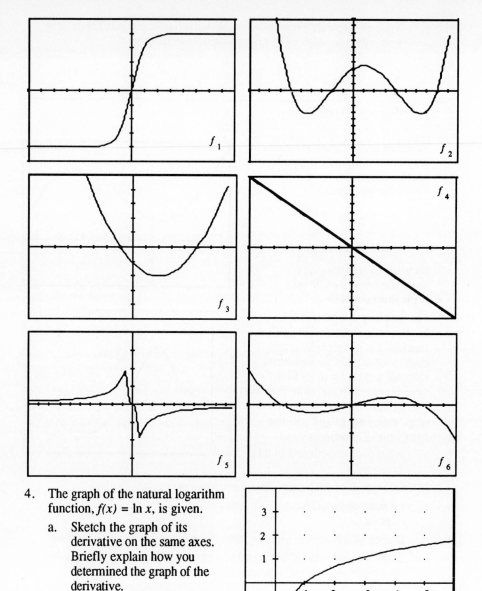

4. The graph of the natural logarithm function, $f(x) = \ln x$, is given.

    a. Sketch the graph of its derivative on the same axes. Briefly explain how you determined the graph of the derivative.

    b. Accurately estimate the derivative of $f(x) = \ln x$ for $x = 2$. Explain your process.

    c. Different from the process chosen in part (b), explain another method that you could use to estimate the derivative of $f(x) = \ln x$ at $x = 2$. (You do not know, nor can you use the function $\frac{d}{dx}\ln(x)$.)

9

Max/Min Problems

9.1 Introduction

One of the most common uses of derivatives in real life is to determine the maxima and/or minima of functions that model real-world situations. Following are four such problems. The first two problems are considered in some depth, allowing consideration of various aspects of each situation. It is left up to the student to determine which questions are appropriate and interesting to consider for the last two situations.

For the following, it is important to (a) state all answers in complete sentences; (b) show all graphs that are used to help solve any problem, labeling the function and the domain and range of the viewing rectangle; and (c) approximate answers correctly to the nearest hundredth.

9.2 The Rain Gutter

A sheet of metal is 200 centimeters long and 25 centimeters wide. It is to be made into a rain gutter by turning up sides of equal height, perpendicular to the sheet. See the figure on the right.

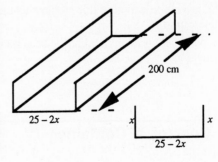

In order to make the rain gutter as efficient as possible, it is desirable to maximize the amount of water the gutter will hold.

1. If sides of 2 cm are bent up, determine the cross-sectional area of the gutter and determine the volume of the gutter.

2. Repeat part (1) if the sides are 3 cm, 6 cm, and 9 cm.

3. Determine an algebraic expression for the cross-sectional area A of the gutter as a function of the height of the turned-up sides x.

4. Draw a *complete* graph of the mathematical function for the cross-sectional area.

5. State the domain of the function in part (4).

6. What portion of the graph is relevant to the physical problem? Why?

7. It has been determined that a volume of at least 12,000 cubic centimeters will be sufficient for most situations in which the gutter will be used.

   a. What cross-sectional area C produces a volume of 12,000 cubic centimeters?

   b. Superimpose the graph of the horizontal line $y = C$ on the graph of the cross-sectional area function from part (4).

   c. What heights for the sides of the gutter would produce volumes of at least 12,000 cubic centimeters? Carefully explain the procedure you used to solve this problem.

8. What height for the turned-up sides will yield the maximum cross-sectional area for the gutter? Explain how this helps determine the maximum volume. Solve this problem graphically. Carefully explain the procedure you used to solve this problem.

9. What is the maximum volume for the gutter?

10. How else might you tackle the problem described in parts (8) and (9)? Describe another method that would lead to an accurate solution.

11. Could derivatives help you solve the problem in part (9)? Explain.

12. State the real-life significance of the x-intercepts (zeros) on the graph of the function.

13. Is any other information included on the graph of the function that has not yet been used?

14. State a modification of this problem that might make it different or more interesting to you.

15. What other questions could have been asked about this (or a similar) problem situation?

9.3 Cost of a Container

A rectangular container is to have an open top and a square base. Assume that the material for the sides of the container costs $3 per square foot and the material for the bottom costs $5 per square foot. The volume of the container is to be 20 cubic feet. (See figure.) In the diagram, x represents the length (and width) of the square base of the container, and h represents the height of the container.

1. Write a formula for the cost of the material for the container in terms of x and h. Let C represent the cost for the material of the container.

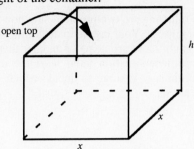

2. Write a formula for the volume of the container in terms of x and h. Let V represent the volume of the container.

3. Since the volume of the container must be 20 cubic feet, use the formula found in part (2) to obtain an equation that gives a relationship between x and h.

4. What is the height of the container and what is the cost of the material for the container if the length of the square base is

    a. 0.5 foot?   b. 1 foot?   c. 2 feet?

    d. 4 feet?   e. 8 feet?   f. 16 feet?

5. Use the results from parts (1) and (3) to write a formula for the cost of the material for the container in terms of x only. What restrictions are there on the possible values of x? Sketch a complete graph of this cost function.

6. The company that manufactures these containers has determined that it can afford a cost of at most $160 per container. What dimensions for the container produce a cost of at most $160? Carefully explain the procedure you used to solve this problem.

7. What dimensions of the container will produce a minimum cost for the material for the container? Solve this problem graphically. Carefully explain the procedure you used to solve this problem. Is your solution consistent with the information obtained in part (4)?

8. What is the minimum cost for the container?

9. How else might you tackle the problem in part (7)? Describe another method that would lead to an accurate solution.

10. Could derivatives help you solve the problem in part (7)? Explain.

11. State a modification of this problem that might make it different or more interesting to you.

12. What other questions could have been asked about this (or a similar) problem situation?

9.4 Additional Max/Min Problems

Following are two problem situations. Questions concerning these applications are not given. Your task is to determine which questions to ask and answer concerning one of these problems or another that you create. Your questions must address the problem situation to at least the same depth as those given in the first two problems. You are to decide which questions are of interest in the situation you choose.

BEAM DEFLECTION

A horizontal support beam in a parking garage is 11 meters long. One-half meter of its left end is built into a concrete wall and one-half meter of its right end is attached to a vertical support, as shown in the figure. The beam is loaded with weight uniformly distributed along its length. As a result, the beam sags downward according to the equation

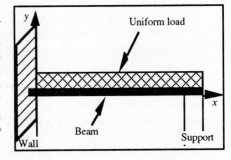

$$y = -x^4 + 25x^3 - 150x^2$$

where x is the number of meters from the wall to a point on the beam, and y is the number of hundredths of a millimeter that the beam sags at that point. (The value for y will be negative.)

COST, REVENUE, AND PROFIT FUNCTIONS

A manufacturer of outdoor furniture has decided to manufacture a new line of high-quality wood swing and slide sets. After studying several factors, the company's mathematicians and financial officers have determined that, for these swing sets, the annual cost (in dollars) to the company of producing x units will be given by the function $C(x) = 0.001x^3 - 0.6x^2 + 132x + 25,000$. They also determined that this function is accurate as long as the company produces fewer than 1000 of these swing sets.

The company plans to sell the swing and slide sets for $600 each. Its officers have determined that it will be able to sell up to 1000 swing sets per year. Thus, the total annual revenue (in dollars) generated by the sale of these swing sets will be $R(x) = 600x$ where x is the number of units manufactured and sold.

10

Linear Approximations

10.1 Introduction

Linear approximation is used to estimate values between and slightly beyond laboratory data sets. If a rate of change is known for a function, and specific data points are known, it is possible to use the rate of change to interpolate in small steps and obtain a reasonable estimate for interpolated values.

Linear approximation is also used to efficiently determine roots of functions. The root-finding method is in turn helpful in solving systems of equations when other algorithms are not available. In the following exercises, the effectiveness of using linear functions to approximate values of non-linear functions will be investigated.

10.2 Using Lines to Approximate Curves

1. a. Determine the equation for the line tangent to the graph of $f(x) = x^2$ through the point (1, 1). Write the equation of the tangent line to $y = f(x)$ through (1, 1) as a function $y = T(x)$.

    b. Sketch the graphs of f and T on the viewing rectangle [0, 2] by [0, 4].[1]

---

[1] Store the formulas for the functions f and T in Y_1 and Y_2 (TI), f_1 and f_2 (CASIO 7700), or store the program consisting of two graph statements, one for the function f and one for the function T (CASIO 7000).

c. Describe the appearance of the tangent line T and the graph of $f(x) = x^2$ for x in the interval $[0.8, 1.2]$.

d. Change the viewing rectangle to $[0.8, 1.2]$ by $[0.6, 1.5]$, and graph the functions again. Compare the graphs of f and the tangent line T on this viewing rectangle.

e. Repeat part (d) using the viewing rectangle $[0.92, 1.08]$ by $[0.8, 1.2]$.

f. How well does the tangent line through $(1, 1)$ approximate the graph of $f(x) = x^2$ on small x-intervals containing $x = 1$? Explain.

2. a. Complete the following table of values to investigate numerically how closely $y = T(x)$ approximates $y = f(x)$ for values of x close to 1.[2]

x	$f(x)$	$T(x)$	$T(x) - f(x)$
0.94			
0.96			
0.98			
1.00			
1.02			
1.04			
1.06			

b. What error is introduced by using $T(1.02)$ to approximate $f(1.02)$? Explain.

c. If $y = T(x)$ is being used to approximate $f(x) = x^2$, to determine how much paint is needed to cover a square region, will the approximation yield a satisfactory estimate of the amount of paint necessary to paint a region with side measurements of $x = 1.02$ meters? Explain.

d. In what situation(s) might more accuracy be necessary? Explain.

3. Repeat an investigation similar to that described in parts (1) and (2) using $f(x) = \sin(x^2)$ near $x = -1.5$.

a. For the graphs of $y = f(x)$ and $y = T(x)$, start with the viewing rectangle $[-2.5, 2.5]$ by $[-1.5, 1.5]$. Choose new viewing rectangles by using the current graph to determine intervals of x and y for which the tangent line and the graph of f nearly coincide.[3]

---

[2] Programs to calculate and display the values $f(x)$, $T(x)$, and the error $E = T(x) - f(x)$ follow (for CASIO 7700 put f in f_1 and T in f_2; for TI, put f in Y_1 and T in Y_2):

CASIO 7000:	CASIO 7700:	TI:	
"X"?→X	'LINAPPROX	Prgm#:LINAPPRX	(cont'd)
x^2→Y◢	"X"?→X	Disp "X"	Disp "E"
2X−L◢	f_1◢	Input X	Y_2-Y_1
"E":Ans−Y	f_2◢	Disp Y_1	Disp Ans
	"E":Ans−f_1	Disp Y_2	

[3] Use program ZOOM (CASIO 7000) or the ZOOM Box option (CASIO 7700 or TI) to simplify this task. Check the range for each successive graph.

b. Create a table similar to that in part (2a) to approximate $f(-1.6), f(-1.5)$ and $f(-1.4)$ using the equation of the tangent line T through $(-1.5, f(-1.5))$. Determine the errors introduced in using this approximation for the values of f.

c. Suggest situations in which

  i.   the estimate for $f(-1.6)$, using the tangent line approximation, is suitable; and

  ii. the estimate for $f(-1.6)$, using the tangent line approximation, is not accurate enough.

4. For a function f that is differentiable on an open interval containing $x = a$, determine the equation for the tangent line $y = T(x)$ through the point $(a, f(a))$.

5. For *small* intervals containing $x = a$, describe the appearance of the graph of $y = f(x)$. Describe the appearance of the graph of $y = T(x)$, the tangent line to the graph of f through the point $(a, f(a))$. How do the graphs of f and T compare on small x-intervals containing $x = a$?

6. The table below lists the world population P (in millions) at 100 year intervals from AD 1500 to 1900, and in ten-year intervals for the 20th century.[4]

Date	Millions	Date	Millions
1500	460	1930	2,070
1600	579	1940	2,295
1700	679	1950	2,515
1800	954	1960	3,019
1900	1,633	1970	3,698
1920	1,862	1980	4,450

a. Graph the data letting $t = 0$ for AD 1500 (so $t = 480$ for 1980).

b. Estimate the instantaneous rate of population growth for 1930 and 1950 from either the table or the graph. Explain both your process and results.

c. Assume that the instantaneous rate of change function for the 20th century data in the table is given by the exponential function

$$R(t) = (0.12016) \cdot 1.01268^t.$$

  i.   Using the method of linear approximations, estimate the world population for 1990.

  ii. Assume that the rate of growth will continue according to the equation $y = R(t)$ above. Use the method of linear approximations in 10-year intervals, first estimating $P(2000)$ from the estimate for $P(1990)$, then estimating $P(2010)$ from the estimate for $P(2000)$, etc., to estimate the world population for the year 2020.

---

[4] Taken from *The Guinness Book of World Records 1991*. New York: Bantam Books, p. 507.

   iii.   It has been estimated that the number of people who have died
          between 40,000 BC and AD 1990 was nearly 60 billion. What
          fraction of the number of persons who have ever lived is the expected
          population for the year 2020?

Related Problems

1. Suppose that the following information is known about a function f.

    $f(0) = 0$ $f(1) = -2$ $f(2) = -4$ $f(3) = 0$
    $f'(0) = 0$ $f'(1) = -3$ $f'(2) = 0$ $f'(3) = 9$
    $f''(0) = -6$ $f''(1) = 0$ $f''(2) = 6$ $f''(3) = 12$

    a. Use a linear approximation for this function to approximate the value of
        $f(1.2)$.
    b. What other values of $y = f(x)$ could be reasonably estimated using the
        given information?

2. Assume that $y = f(x)$ and $y = h(x)$ are differentiable. Also assume that $f(2) = 1$,
    $f'(2) = 3$, $h(-4) = 2$, and $h'(-4) = -1$. Find the following:

    a. the equation of the line tangent to the graph of $y = h(x)$ when $x = -4$
    b. an estimate of $h(-3.8)$ using a linear approximation
    c. an estimate of $f(2.1)$ using a linear approximation
    d. the equation of the line tangent to the graph of $g(x) = f(h(x))$ when $x = -4$

3. a. Sketch the graph of $f(x) = x^3 - 2x + 1$ on a viewing rectangle $[-5, 5]$ by
        $[-10, 10]$.
    b. Find the tangent line T to the graph of f through the point $(-2, f(-2))$.
        Overlay the graph of T.
    c. Describe the appearance of the tangent line T as compared to the graph of f
        on this window.
    d. Graph f and the tangent line T on a viewing rectangle $[-3, -1]$ by
        $[-8, 2]$. Describe the appearance of the tangent line T as compared to the
        graph of f on this window.
    e. Would the value of x where $T(x) = 0$ give a good approximation for the
        value of x where $f(x) = 0$? Explain.
    f. If the point of tangency $(a, f(a))$ is chosen close to an x-intercept for a
        function f, how might the tangent line through $(a, f(a))$ compare to the
        graph of f on an interval containing the x-intercept? Explain.
    g. Explain how tangent line approximations might be used to estimate a root
        for a differentiable function f.

10.3 Newton's Method

Preliminaries

When there are no methods that will give an exact solution of an equation $f(x) = 0$, numerical methods are used to approximate the solution. Newton's Method uses the instantaneous rate of change of a function to speed the process of determining a root. To use this method, you first need an approximation a of the root r. Graphically, Newton's Method obtains the next approximation b by using a

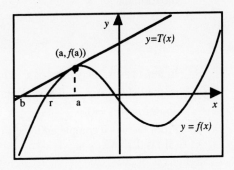

tangent line to the graph of the function f at the point $(a, f(a))$ as illustrated in the figure above.

The solution of $f(x) = 0$ corresponds to an x-intercept r of the graph of f. (The exact value of r is not known.) If a is an approximation of the solution, then the tangent line to the graph of f at the point $(a, f(a))$ uses the rate of change of f at $x = a$ to provide another approximation b of the root.

1. Determine the equation of the line tangent to the graph of f at $(a, f(a))$. Write this equation as a function $y = T(x)$.

2. This tangent line is the linear approximation for the function f at $x = a$ (see Section 10.2). Newton's Method uses the function T (tangent line) as an approximation for the function f. Describe the relationship between f and T on a small interval about $x = a$ to suggest why the tangent line might be helpful in efficiently determining a root.

3. Since it is not possible to solve the equation $f(x) = 0$, the equation $T(x) = 0$ will be solved to obtain the next approximation for the solution of $f(x) = 0$.

    a. In the figure above, locate the point on the graph of $y = T(x)$ for which $T(x) = 0$.

    b. For the equation of $y = T(x)$ found in part (1), determine an expression for the x-intercept of the tangent line, $x = b$ (where $T(b) = 0$).

4. The x-intercept of the tangent line serves as the next approximation for the solution of $f(x) = 0$. If this x-intercept is denoted by b, the procedure is repeated by finding the x-intercept of the line tangent to the graph of f at the point $(b, f(b))$. Locate the point $(b, f(b))$ on the graph above. Sketch a line tangent to the graph of f through the point $(b, f(b))$. Locate the x-intercept of this second tangent line. Is it a better approximation for the root than $x = b$? Explain.

5. This procedure is usually stated in recursive form as follows.

a. Start with an initial approximation x_0 for the solution of $f(x) = 0$.

b. To go from the nth approximation x_n to the next approximation x_{n+1}, use the formula

$$x_{n+1} = x_n - \frac{f(x_n)}{f'(x_n)}$$

In most cases, the values of x_n will converge rapidly to the solution r and so the values of x_n will give decimal approximations to the solution of the equation $f(x) = 0$.[5]

c. From the work completed in part (3b), explain how the formula in part (5b) is derived.

Investigation

1. The graphical nature of Newton's Method is investigated as follows:[6]

a. Let $f(x) = x^2 - 10$. (A positive solution of the equation $f(x) = x^2 - 10 = 0$ provides a decimal approximation for the square root of 10.) Sketch the graph of $f(x) = x^2 - 10$ using the viewing rectangle [0, 6] by [–10, 20].

b. Use Newton's Method with an initial approximation of 1 for the equation $x^2 - 10 = 0$.

i. Start by superimposing the graph of the line tangent on the graph of f at the point (1, –9) and determining the x-intercept of this tangent line. This will be the next approximation of the solution of the equation.

ii. Repeat this process at least two more times to generate successive approximations of the solution of the equation. What solution is obtained?

c. Clear the graphics screen and repeat parts (a) and (b) using an initial approximation of A = 5. Did this give the same solution as before?

2. Solve the equation $\cos x = 0.5x$.

a. Rewrite the equation in the form $\cos x - 0.5x = 0$. Then use Newton's Method to determine a root for the function $f(x) = \cos x - 0.5x$.

b. Sketch the graph on the viewing rectangle [–5, 5] by [–5, 5].

---

[5] Programs NEWTON.G and NEWTON are used in this investigation. The initial approximation for the solution of the equation is denoted by A.

[6] Use NEWTON.G (CASIO and TI) to sketch these tangent lines and find their x-intercepts, or use a program on a computer.

c.   Use Newton's Method with an initial approximation of 0 to solve this equation . On the graph of the function, draw at least two tangent lines used in Newton's Method. What is the solution?

3.   Repeat part (2) using an initial approximation of 3.

a.   Record each approximation. Stop after drawing three tangent lines.

b.   Does it seem likely that the approximations will converge to the zero of the function? Explain.

c.   Was 3 a good choice for the initial approximation? Explain in terms of the graph of f and the slope of the tangent line at $(3, f(3))$. (Actually, if enough repetitions are performed, Newton's Method will converge to the root. However, this takes a great deal of patience.)

4.   The function $f(x) = \sin x$ has a zero at $x = \pi$. A decimal approximation for π is 3.14159265359.

a.   Sketch the graph of $f(x) = \sin x$ on the viewing rectangle [–15, 15] by [–2, 2].

b.   Use Newton's Method with each of the following initial approximations. In each case, draw at least two tangent lines used in Newton's Method.

i.   3   ii.   1.9   iii.   1.8   iv.   1.5

c.   Why were such different results obtained for these four different initial approximations?

5.   Let $f(x) = 3x^3 - 24x^2 + 60x - 46$. Notice that $f(2) = 2$ and $f(3) = -1$. Thus, according to the Intermediate Value Theorem, the function f has a zero between 2 and 3.

a.   Sketch the graph of the function f on the viewing rectangle [0, 5] by [–10, 10].

b.   Use Newton's Method with each of the following initial approximations. In each case, draw at least two tangent lines used in Newton's Method.

i.   2   ii.   2.1   iii.   2.2   iv.   1.9

What happened in part (5b i)? Why did it happen? Why were different results obtained in parts (5b ii), (5b iii), and (5b iv)?

6.   Describe situations in which Newton's Method may fail to converge to the desired root of an equation. Use properties of the graph of the function and properties of the derivative of the function in your description.

7.   How might Newton's Method be used to determine maxima or minima for a function $y = f(x)$? Give two examples using the Max/Min problems in Chapter 9 of this manual, or the Diving Board problem found in the appendices.

10.4 Related Problems

Exercise (1) assumes use of a Newton's Method program.

1. Solve the equation $\sin x = 0.5x$ by rewriting the equation as
    $\sin x - 0.5x = 0$.

    a. Sketch a graph of the function $f(x) = \sin x - 0.5x$. Use this graph to
        obtain an initial approximation of the solution of the equation and then
        use Newton's Method.

    b. What is the solution? Check the solution.[7] Explain the result.

    c. Find all solutions of the equation $x^3 - 4x^2 + x + 3 = 0$ and check each
        solution.

    d. Find all solutions of the equation $3 \cdot \sin x = \dfrac{x^2}{4}$ and check each solution.

    e. Find all points of intersection of the graphs of the functions
        $f(x) = 4x^3 - 12x^2$ and $g(x) = 4x - 24$.

2. Sketch the graph of the function $h(x) = x \cos x + \sin 2x$ on the interval
    $[0, 5]$. Find all relative maxima and minima of the function on the interval
    $[0, 5]$.

3. The graph of a function f that
    shows a solution to the equation
    $f(x) = 0$ at $x = r$ is provided in the
    figure at the right. Assume that r
    cannot be found by analytic
    methods and that Newton's Method
    will be used to approximate r.

    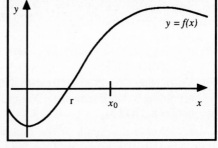

    The graph shows a first
    approximation x_0 for r (to be used
    for Newton's Method). Give a
    geometric explanation of the procedure that Newton's Method uses to find the
    next approximation (x_1) for r, and show (on the graph) how the next two
    approximations $(x_1$ and $x_2)$ can be found.

4. The graph of a function $y = f(x)$ is
    provided in the figure at the right.
    Assume that $f(1) = 1.12$ and that
    $f\,'(1) = -3.84$.

    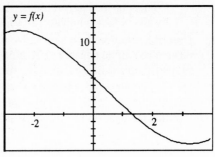

    a. Find the equation of the line
        tangent to the graph of
        $y = f(x)$ at the point $(1, 1.12)$.

    b. Using $x = 1$ as the initial
        approximation for the solution
        of the equation $f(x) = 0$,
        determine the next approximation for that solution given by Newton's
        Method.

---

[7] For CASIO and TI graphics calculator users, after using the program NEWTON, store
the value of R in memory location X and evaluate $f(X)$. To evaluate $f(X)$, run Program
0 (CASIO 7000) or evaluate f_1 (CASIO 7700) or Y_1 (TI).

11

Interpreting the
Second Derivative

11.1 Introduction

To investigate the meaning of the second derivative, we explore the relationship between velocity and acceleration with respect to the concavity of the graph of a function. In addition, the relationship between the graph of the function and the graph of the first derivative of the function as these relate to the concavity of the function are explored.

Roughly, the graph of a function f is concave up over an interval if any secant line within that interval lies above the graph of the function. The function f is concave down over the interval if secant lines to the graph of f lie below the graph of the function.

For this investigation, the following functions and viewing rectangles will be used.

function	$y = f(x)$	x-interval	y-interval
1.	$f(x) = (x + .5)^2 - 2$	$[-0.5, 2.5]$	$[-3, 5]$
2.	$f(x) = 2\sqrt{x} - 2$	$[0, 5]$	$[-2, 3]$
3.	$f(x) = (x - 2)^3 - 2$	$[0, 5]$	$[-10, 15]$
4.	$f(x) = \sqrt[3]{x}$	$[-5, 5]$	$[-2, 2]$

11.2 Velocity Versus Acceleration[1]

Think about an object whose vertical position changes while its horizontal position is held constant. An elevator or a spider climbing a wall might be two such objects.

1.   Assume that each of the functions listed in Section 11.1 represents the position of an object with respect to time. Complete the following for function 1, $f(x) = (x + .5)^2 - 2$.

    a.   Graph the function.[2]

    b.   Imagine that an object is climbing vertically and that its position at any point in time is given by the equation listed. How is the object climbing? Slowly, quickly, not moving, first slowly then more quickly, etc.?

    c.   Describe the change in velocity as the object climbs.

    d.   How does the velocity of the object correspond to the concavity of the function?

    e.   Repeat parts (1a) through (1d) with functions 2 through 4.

2.   Answer the following using the information from the observations made in part (1).

    a.   How is the object moving over intervals when the graph is concave down? How is its velocity changing?

    b.   How is the object moving over intervals when the graph is concave up? How is its velocity changing?

    c.   Is the graph concave up or concave down over intervals for which the velocity of the object is increasing? Explain.

    d.   Is the graph concave up or concave down over intervals for which the velocity of the object is decreasing? Explain.

3.   a.   What is the relationship between the velocity and the acceleration of an object in calculus terms?

    b.   When the velocity of an object is increasing, the object has positive acceleration. In calculus terms, from the position function for the object, when is an object accelerating positively?

    c.   When the velocity of an object is decreasing, the object has negative acceleration. In calculus terms, from the position function for the object, when is an object accelerating negatively?

---

[1]   This section is best illustrated through the use of a program that illustrates the change in y as x stays constant while simultaneously sketching the function $y = f(x)$. The program SPIDER is provided in the appropriate appendix for graphing calculator users.

[2]   Use the program SPIDER found in the appropriate appendix to illustrate the vertical movement as the distance-versus-time graph is also being sketched.

11.3 Graphical Interpretation of the Second Derivative

Note that all of the functions listed in Section 11.1 are increasing functions. Intuition about velocity (directional speed) might interfere with your understanding when we discuss decreasing functions. Section 11.3 is designed to look at the relationship between functions and their first and second derivatives graphically. It is left to the student to put the information from Sections 11.2 and 11.3 together and relate the graph of f with information about first and second derivatives as well as velocity and acceleration.

1. Using the corresponding axes below, sketch the graphs of the functions 1 through 4 (listed in Section 11.1) for the indicated viewing rectangles. Label the x and y scales on the axes.

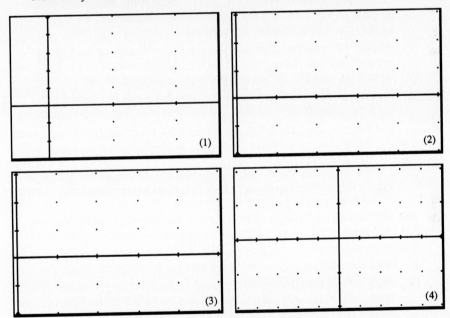

2. Complete the following for each of functions 1 through 4.

    a. Determine intervals over which f is concave up.

    b. Determine intervals over which f is concave down.

    c. Sketch several tangent lines along the curve. Observe the steepness of successive tangent lines as x increases over the interval given.[3]

---

[3] Graphing calculator users might find the program TANGENT (in appropriate appendix) useful in completing part of this investigation. TANGENT sketches the

     i.   What can be said about the steepness of successive tangent lines over intervals in which f is concave up?

    ii.   What can be said about the steepness of successive tangent lines over intervals in which f is concave down?

d.   The function that gives the slopes of tangents lines for a function f is the first derivative of f.

     i.   The first derivative of a function f is _____ (increasing/decreasing) when f is concave up.

    ii.   The first derivative of a function f is _____ (increasing/decreasing) when f is concave down.

3. a.   What is the sign of the first derivative $y = g'(x)$ of a function $y = g(x)$ when g is increasing?

    b.   What is the sign of the first derivative $y = g'(x)$ of a function $y = g(x)$ when g is decreasing?

    c.   From part (2), when f is concave up, is f' increasing, decreasing, or neither? Explain. Since $y = f''(x)$ is the (first) derivative of $y = f'(x)$, what is the sign of f'' when the graph of f is concave up?

    d.   When $y = f(x)$ is concave down, is $y = f'(x)$ increasing, decreasing, or neither? Explain. Since $y = f''(x)$ is the (first) derivative of $y = f'(x)$, what is the sign of f'' when the graph of f is concave down?

    e.   For each of the functions 1 through 4, determine an expression for $y = f'(x)$. Sketch the graphs of $y = f(x)$ and $y = f'(x)$ on the same axes.

     i.   For intervals over which f is concave up, is the graph of f' increasing, decreasing, or neither? How does this information compare to that found in part (2)?

    ii.   For intervals over which f is concave down, is the graph of f' increasing, decreasing or neither? How does this information compare to that found in part (2)?

4.   The four functions sketched in part (1) are all increasing functions. Suppose that f is a function that is decreasing on an interval and is concave down over that interval.

    a.   Draw the graph of such a function. Label the scales on the axes.

    b.   Draw several tangent lines on the graph of the function f. Estimate the values of the slopes of these tangents. Are the values of the slopes of these tangent lines increasing, decreasing, or neither?

    c.   Is $y = f'(x)$ increasing, decreasing, or neither? Explain.

    d.   What is the sign of $y = f'(x)$ on this interval where the function $y = f(x)$ is decreasing?

    e.   What is the sign of $y = f''(x)$ on this interval where the function $y = f(x)$ is concave down? How does your answer compare to those in parts (2) and (3)?

---

graph of $y = f(x)$, then sketches tangents along the curve while it plots the values of the slopes of tangents, sketching an approximate curve of $y = f'(x)$.

f. Sketch the graph $y = f'(x)$ on the same axes as in part (4a). Is the graph of f' increasing, decreasing or neither? How does this information compare to that found in part (4e)?

5. Suppose that f is a function that is decreasing on an interval and is concave up on that interval.

   a. Draw the graph of such a function. Label the scales on the axes.

   b. Draw several tangent lines on the graph of the function f. Estimate the values of the slopes of these tangents. Are the values of the slopes of these tangent lines increasing, decreasing, or neither?

   c. Is $y = f'(x)$ increasing, decreasing, or neither? Explain.

   d. What is the sign of $y = f'(x)$ on this interval where the function $y = f(x)$ is decreasing?

   e. What is the sign of $y = f''(x)$ on this interval where the function $y = f(x)$ is concave up? How does your answer compare to that in parts (2) and (3)?

   f. Sketch the graph $y = f'(x)$ on the same axes as in part (5a). Is the graph of f' increasing, decreasing, or neither? How does this information compare to that found in part (5e)?

6. The following functions will be used to further explore graphically the relationships between a function, its first derivative, and its second derivative.

   i. $f(x) = 4 - x^2$ ii. $f(x) = x^2 - 4$ iii. $f(x) = (x - 1)^3$

   For each of the functions (i) through (iii), complete the following.

   a. Sketch the graphs of the function $y = f(x)$, its first derivative $y = f'(x)$, and its second derivative $y = f''(x)$ on the same axes.

   b. Determine the properties of the graph of the first derivative function when the graph of the function is concave up or concave down. Is the first derivative positive or negative, increasing or decreasing, etc.? Explain.

   c. What is the sign of the second derivative when the graph of the function is concave up or concave down?

   d. Describe the appearance of the graph of $y = f(x)$ where its first derivative is equal to zero.

   e. Describe the appearance of the graph of $y = f(x)$ where its first derivative and its second derivative are equal to zero.

   f. Describe the appearance of the graph of $y = f(x)$ where its first derivative is not equal to zero, but its second derivative is equal to zero.

   g. If a function is given symbolically, is it possible to determine where the function will be concave up or concave down simply by using information from its derivative functions? Explain.

   h. How might the second derivative be used to help determine whether or not a critical point yields a maximum or minimum for the function $y = f(x)$?

11.4 Related Problems

1. Refer to the functions whose graphs are given below to answer the following
    questions. Provide a brief explanation of your responses.

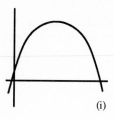
(i)

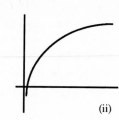
(ii)

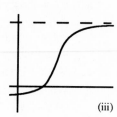
(iii)

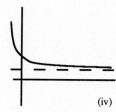
(iv)

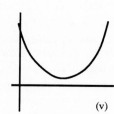
(v)

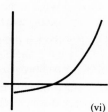
(vi)

   a. Which functions have a positive first derivative for all x?
   b. Which functions have a negative first derivative for all x?
   c. Which functions have a positive second derivative for all x?
   d. Which functions have a negative second derivative for all x?

2. The graph of the derivative
    $y = f'(x)$ of a polynomial
    function $y = f(x)$ is shown on the
    right.

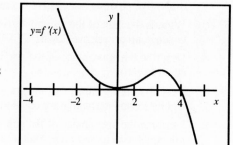

   a. Determine the largest intervals
       over which the function is:
       i.   increasing,
       ii.   decreasing,
       iii. concave up, and
       iv.   concave down.

   b. For $x > 0$, if $y = f(x)$ represents the distance of an object above the ground
       with respect to time:
       i.   What does $y = f'(x)$ represent?
       ii.   At what point is the object furthest from the ground? Explain.
       iii. What does the maximum near $x = 3$ on $y = f'(x)$ represent in terms of
             the position $y = f(x)$ of the function? Explain.

3. The function $y = f''(x)$ is graphed in the figure on the right. Sketch the graph of $y = f'(x)$ given $f'(0) = 3$. Sketch a possible graph of $y = f(x)$. Briefly explain the process you used to determine the graph of f.

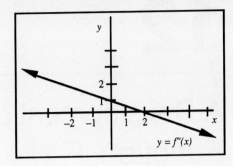

4. The graphs of $y = f'(x)$ and $y = f''(x)$ are provided in the figure at the right. On the same axes, sketch the graph of $y = f(x)$ which could have f' and f'' as its first and second derivatives respectively.

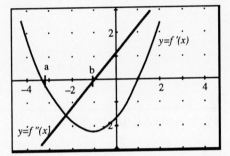

5. The graph of $y = f'(x)$ is provided at the right. Assume that the function $y = f(x)$ represents the number of sales of an innovative calculus workbook in its first several years. **Note that this is the graph of the derivative of f, not the graph of the function f itself.**

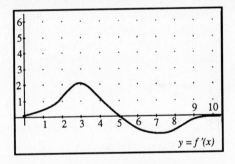

   a. What does f' represent in this situation?

   b. During which year(s) were the sales increasing most rapidly? Explain.

   c. A point on the graph of a differentiable function where the concavity changes is called an inflection point. Locate an inflection point for the graph of f (not for f'). What does this inflection point represent in terms of the sales of the calculus workbook?

12

A Development of the Function $F(x) = e^x$

12.1 Introduction

Using the laws of exponents, and differential calculus, properties of the function $y = e^x$ will be developed. This approach is different from the usual approach taken in calculus texts, in that it does not use the definition of the natural logarithm function. It is provided to give students more experience with using calculus to develop properties of unfamiliar functions, and to assist students in becoming more familiar with exponential functions.

12.2 Discovering Properties of Exponential Functions

1. Let $f(x) = 2^x$, $g(x) = 3^x$, and $k(x) = 4^x$.
   a. Sketch the graphs of these three functions on the same set of axes using the viewing rectangle [–2, 3] by [0, 20].
   b. Repeat part (a) using the viewing rectangle [–1, 1] by [0, 4].
   c. For each function
      i.   state the domain and range, and
      ii. determine all of the x and y intercepts.
   d. Compare the graphs of f, g, and k. Consider three cases: x in $(-\infty, 0)$, $x = 0$, and x in $(0, \infty)$. For each case, discuss the following properties of the functions f, g, and k as these functions compare with each other.

      i.   Rate of change

     ii.   Concavity

    iii.   Relative magnitudes of 2^x, 3^x, and 4^x for each x

  e.   i.   For $0 < b < 4$, for what values of x is $b^x > 4^x$?

     ii.   For $b > 4$, for what values of x is $4^x < b^x$?

  f.   For which graph (f, g, or k) does it appear that the slope of the tangent line at $(0, 1)$ has the greatest value? the smallest value?

2.   For the function $f(x) = 2^x$, consider the slopes of the secant lines that pass through the points $(0, 1)$ and $(h, f(h))$.

  a.   Write an expression for the slope m of the secant line that passes through the points $(0, 1)$ and $(h, f(h))$. Remember, $f(h) = 2^h$.

  b.   Complete the following table showing the values of h and the corresponding slopes m of the secant lines.[1]

h	m	−h	m
1		−1	
0.5		−0.5	
0.25		−0.25	
0.125		−0.125	
0.0625		−0.0625	
0.03125		−0.03125	
0.015625		−0.015625	
0.0078125		−0.0078125	

  c.   What conclusions can you draw about the slope of the line tangent to the graph of $f(x) = 2^x$ at $(0, 1)$, based on the information in part (2b)? Explain.

  d.   i.   What is an approximate value of $f'(0)$? Explain.

     ii.   Is it possible that $f'(0) = 1$? Explain.

3.   For the function $g(x) = 3^x$, consider the slopes of the secant lines that pass through the points $(0, 1)$ and $(h, g(h))$.

  a.   Write an expression for the slope m of the secant line that passes through the points $(0, 1)$ and $(h, g(h))$. Remember, $g(h) = 3^h$.

  b.   Complete the following table showing the values of h and the corresponding slopes m of the secant lines.[2]

---

[1] Use SECANT or BEHAVIOR to complete this table. For BEHAVIOR, the expression for the slope of a secant must be written as a function of X.

[2] Use SECANT or BEHAVIOR to complete this table. For BEHAVIOR, the expression for the slope of a secant must be written as a function of X.

h	m	−h	m
1		−1	
0.5		−0.5	
0.25		−0.25	
0.125		−0.125	
0.0625		−0.0625	
0.03125		−0.03125	
0.015625		−0.015625	
0.0078125		−0.0078125	

c. What conclusions can you draw about the slope of the line tangent to the graph of $g(x) = 3^x$ at $(0, 1)$, based on the information in part (3b)? Explain.

d. i.   What is an approximate value of $g'(0)$? Explain.

ii. Is it possible that $g'(0) = 1$? Explain.

e. Compare the results of parts (2d) and (3d). We are interested in determining the value of b so that, for the function $F(x) = b^x$, $F'(0) = 1$. On what interval would you expect to find this number b?

4. Let $F(x) = b^x$ $(b > 0)$.

a. Use the definition of the derivative to write an expression for $y = F'(x)$.

b. Use part (4a) to write an expression for $F'(0)$.

c. Use the laws of exponents, properties of limits, and the expressions found in parts (4a) and (4b) to write $y = F'(x)$ in terms of $F'(0)$. What does this expression tell you about the derivative of $F(x) = b^x$ as compared to the function $F(x) = b^x$?

5. a. For which values of b is $y = F(x)$ increasing? Decreasing? Explain.

b. Does $F(x) = b^x$ have a maximum or minimum? Explain.

c. Use part (4c) to find an expression for $F''(x) = \dfrac{d}{dx}(F'(x))$. Note: It is not necessary to use the definition of the derivative this time. Also, remember that $F'(0)$ is a constant.

d. Use the results of part (5c) to determine the concavity of the graph of $F(x) = b^x$ over its entire domain. Is the concavity of $F(x) = b^x$ $(b > 0)$ dependent on the value of b?

6. The results in parts (2) and (3) suggest that the slope of the line tangent to the graph of $f(x) = 2^x$ at $(0, 1)$ is less than 1 and the slope of the line tangent to the graph of $g(x) = 3^x$ at $(0, 1)$ is greater than 1. It seems reasonable to conclude that there is some number b between 2 and 3 such that the line tangent to the graph of $F(x) = b^x$ has slope equal to 1. In fact, the number e can be defined as the real number between 2 and 3 such that if $F(x) = e^x$, then $F'(0) = 1$.

a. Use the expression found in part (4c) to write an expression for $y = F'(x)$ when $F(x) = e^x$ and $F'(0) = 1$.

b. How does $F(x) = e^x$ compare with $\frac{d}{dx}(F(x))$?

12.3 Numerical Approximations of *e*

Complete Section 12.2 of this investigation before beginning this section. Numerical approximations for the number *e* will be obtained using a calculator and the definition of *e* found in Section 12.2 part (6).

1. a. For small choices of h, the slope of the secant line $m = \dfrac{f(c+h) - f(c)}{h}$ can be used to approximate the slope of the tangent line. Explain why this is true.

b. For the function $F(x) = b^x$, and for a value of h close to 0, write an expression to approximate $F'(0)$ without using limits. Write this expression as a function of x, $y = G(x)$.

c. With h = 0.0001, sketch the graphs of $y = G(x)$ and $y = 1$ on the same axes. Start with the viewing rectangle [0, 4] by [–2, 2]. Zoom in to approximate the x-coordinate of the point of intersection of the graphs to the nearest thousandth.

d. What is the significance of the x-coordinate of the point of intersection of the graphs in part (1c)? (Recall your work from Section 12.2.)

e. Create a table similar to that begun below to numerically determine the point of intersection of the two graphs in part (1c). The value $x = b$ found in this exercise is the value for which $G(b) = 1$. Let h = 0.0001.

b	$G(\mathbf{b})$
2.5	0.91633
2.6	0.95556
2.7	0.99330
2.8	1.02967

For each choice of b, compare $G(b)$ with $y = 1$. The table shows that b is between 2.7 and 2.8. Continue refining the choice of b until the estimate for b is accurate to the nearest thousandth.[3] Carefully explain the procedure you used.

---

[3] To assist in this numeric investigation, store the formula in terms of x in f_1 (CASIO 7700) or in Y_1 (TI) and evaluate for various choices of X. On the CASIO 7000, enter

    f. How does your result in part (d) compare with the approximation for e on your calculator or in the course text? Explain your results.

    g. Does this investigation provide an approximation for the number e? Why or why not? (Use more than the comparison in part (1f) to justify your findings.)

2. a. Sketch a graph of the function $f(x) = \left(1 + \dfrac{1}{x}\right)^x$ on the viewing rectangle [0, 20] by [0, 3]. Superimpose the graph of the horizontal line $y = e$.

    b. It can be shown that the line $y = e$ is a horizontal asymptote for the graph of the function f. Using the definition of a horizontal asymptote, write e as a limit.

    c. Complete the table below, extending it as far as necessary to obtain an estimate for e that is accurate to the nearest thousandth.[4] How does this value compare to that obtained in part (1e)? How does this value compare with the value of e from your calculator or course text?

x	$\left(1 + \dfrac{1}{x}\right)^x$
100	2.70481
200	2.71152
400	
800	

---

the following program : ?→B:(B $\boxed{x^y}$ H − 1) ÷ H. Store 0.0001 in memory location H (CASIO and TI) before evaluating the expression or running the program.

[4] On the CASIO and TI, use BEHAVIOR with H+2→H replaced by the line H × 2→H. Let A = 0 and H = 100.

13

Determining Distance from Velocity

13.1 Introduction

The velocity function can obtained from the first derivative of a distance function. When the velocity function is provided, how might the distance function be determined? This exploration considers the relationship between a distance function and a velocity function when the velocity function is given.

13.2 Andy's Travel

1. Suppose that Andy travels at a constant rate of 60 mph for 3 hours.

    a. What is Andy's velocity at any moment during the time period from $t = 0$ to $t = 3$?

    b. Write an algebraic expression for the velocity function that models this situation.

    c. How far does Andy travel in these three hours?

    d. Sketch the graph of the velocity function for t in [0, 3] on the axes provided above.

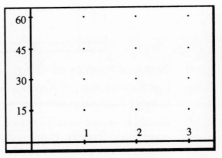

e.   Sketch a vertical line from the *t*-axis to the graph of the velocity function through *t* = 3. What kind of geometric figure is sketched? How does the total distance Andy traveled relate to the graph of his velocity function? Explain.

2.   Suppose Andy traveled 60 mph for 2 hours, then traveled for 1 more hour at 45 mph.

a.   Sketch the graph of the velocity function for *t* in [0, 3], labeling both the axes and the graph of the function.

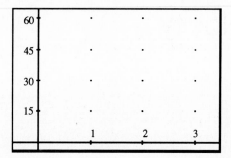

b.   What is Andy's velocity at any moment during the first 2 hours? At any moment during the last hour?

c.   Write an expression that gives the velocity function that models this situation. (Use a piecewise-defined function.)

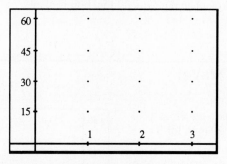

d.   How far does Andy travel in these 3 hours?

e.   Can the total distance Andy traveled be determined from the graph of his velocity function alone? Explain.

f.   Describe Andy's velocity at exactly 2 hours after his travel begins. Is this reasonable? Explain. If this situation is not reasonable, sketch a graph which models a more reasonable situation.[1]

---

[1] It is important that you understand how your numerical and graphical results relate to each other. Discuss your findings with other members of your class. In particular, share your graphs drawn as a result of part (2f) and discuss the answer to part (2e) in light of your new graphs.

3. Following is a more realistic rendering of the situation described in part (2).

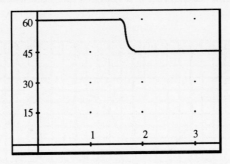

a. Is it possible to determine the distance Andy traveled alone, from this graph? Explain.

b. Suggest methods that could be used to determine the distance Andy traveled from the sketch of the graph.

c. Use one of the methods suggested in (3b) to obtain a rough estimate of the distance Andy traveled.

d. How might a more accurate estimate of the distance Andy traveled be obtained? Find this better estimate.

4. Suppose Andy's velocity is decreasing at a constant rate from 60 mph to 0 mph over a period of 3 minutes.

a. Sketch the graph of the velocity function for t in [0, 3]. Label the axes and the graph of the function.

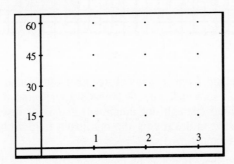

b. What is Andy's velocity at $t = 1$ minutes and at $t = 2$ minutes?

c. Write an algebraic expression for the velocity function that models this situation.

d. Look carefully at the velocity graph sketched in part (4a). Can the total distance Andy traveled be determined from the graph of his velocity function alone? Explain.

e. How far does Andy travel in these 3 minutes?

5. The horizontal **velocity** of a certain type of rocket is modeled by the following graph. Note that this is a velocity function, **not** a position-versus-time function.

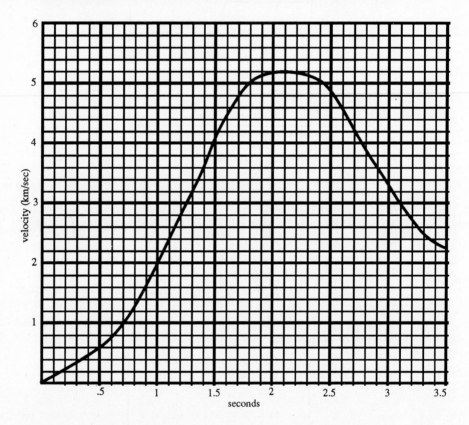

The rocket is launched from a law enforcement compound intended to hit a marijuana field that is surrounded by the archaeological ruins of an ancient village. The village is fortified (though not inhabited) and is of historical importance. Officials are interested in destroying the marijuana field without disturbing the ruins. See the picture below.

Note: Each large square in the graph represents: 1 km/sec × $\overline{0.5}$ sec = 0.5 km

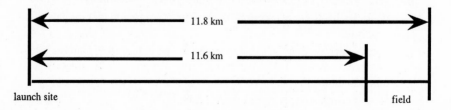

a. Using the graph of the velocity function, how can the distance the rocket will travel be determined? Give a graphical interpretation of distance for this velocity function.

b. From the graph of the velocity function, determine whether the rocket will hit the field, land short of the field (thereby causing damage to the ruins), or land beyond the field (also causing damage).

Note: You will need to determine both a low estimate for the distance traveled and a high estimate. The boundary of regions used in the estimation cannot overlap the velocity graph.

c. Discuss your results with other members of your class. How can a better approximation of the distance traveled be achieved? Explain.

d. Discuss the problem of how to find the exact distance traveled when given the graph of the velocity function. How does finding overestimates and underestimates for the distance traveled help in determining the exact distance traveled? How can these over- and underestimates be refined?

6. a. Graph $y = 10x^2$ on [0, 2] by [0, 45].[2] Notice that a region under the curve of $f(x) = 10x^2$ on [0,2] is enclosed by the lines $y = 0$ and $x = 2$, and by the graph of $y = f(x)$.

b. i. Without changing the y-interval, graph $y = f(x)$ for x in [1, 2].[3]

ii. Graph $y = f(1)$ and $y = f(2)$.[4] Sketch the results.

c. Change the domain to [1.5, 2], graph $f(x) = 10x^2$, then sketch $y = f(Xmin)$ and $y = f(Xmax)$ for the x-interval [1.5, 2].

d. Repeat part (6c) for x in [1.9, 2].

e. Repeat part (6c) for x in [1.99, 2].

f. Repeat part (6c) for x in [1.999, 2].

g. As h decreases in size, describe the appearance of the graph of $f(x) = 10x^2$ from $x = 2 - h$, to $x = 2$.

h. Describe how the line $y = f(Xmin)$ approximates $y = f(x)$ on the interval [2 – h, 2] for small h.

i. Describe how the line $y = f(Xmax)$ approximates $y = f(x)$ on the interval [2 – h, 2] for small h.

j. How might this information be useful in determining the distance traveled by an object whose velocity is given by $f(x) = 10x^2$ between $x = 2 - h$ and $x = 2$?

k. Discuss the results obtained in parts (6a) through (6j) with other members of your class before going on to the Section 13.3.

---

[2] Use program 1 (CASIO 7000) , f$_1$ (CASIO 7700), or Y$_1$ (TI) to graph this function to save time later.
[3] Reset Xmin each time to graph f on the new domain.
[4] Program AREA is provided in the appendices to assist in this exploration.

13.3 Numerical Approximations

1. By now you have discovered that the distance traveled by an object whose velocity function f is given graphically can be determined by finding the area under the curve of f. Using a numerical approximation program (with left and right endpoint approximating options) on a computer or calculator, complete the following to approximate the area under the graph of $f(x) = x$ over an interval $[a, b]$, using lower rectangles L_n then upper rectangles U_n, for n subintervals.[5]

    a. Sketch $y = f(x)$ for x in $[0, 5]$.

    b. Suppose $y = f(x)$ is a velocity function. Determine the distance traveled for x in $[0, 5]$ by finding the area under $y = f(x)$ between $x = 0$ and $x = 5$ using a well-known geometric formula.

    c. Using a numerical approximation program on a computer or calculator, determine an estimate of the area under the curve of $f(x) = x$ using rectangles whose height is determined by the function value of the left endpoint of each of 4 equal subintervals. Sketch $y = f(x)$ and the rectangles found. Are these lower or upper rectangles?

    d. Determine an estimate of the area under the curve of $y = f(x)$ using lower rectangles for n = 8, 16, 32, and 64 equal subintervals respectively. Record the results you obtain in the table below.

n subintervals	Area under $y = f(x)$ estimated by	
	lower rectangles, L_n	upper rectangles, U_n
4		
8		
16		
32		
64		

    e. Compare subsequent estimates. As n gets large, is the estimate for the area under the curve improving? Explain.

    f. Repeat parts (1b), (1c), and (1d) using rectangles whose height is determined by the function value of the right endpoint of each of the n subintervals.

    g. Does one numeric method of determining the area under the curve provide a better estimate for the area than another numeric method for this function? Explain.

2. Approximate the area under the curve of $f(x) = 10x^2$ over the interval $[0, 2]$ using lower rectangles L_n then upper rectangles U_n for n subintervals.

---

[5] Use the program RIEMANN on the CASIO or TI to complete this investigation.

a. Does the sum of the areas of the lower rectangles give an underestimate or an overestimate for the area under the curve of $f(x) = 10x^2$ on the interval $[0, 2]$? Explain.

b. Do upper rectangles give an underestimate or an overestimate for the area under the curve of $f(x) = 10x^2$? Explain.

c. How can a better approximation for the area under $y = f(x)$ be obtained?

d. List the values of successive underestimates and overestimates in the following table.

n subintervals	Area under $y = f(x)$ estimated by	
	lower rectangles, L_n	upper rectangles, U_n
4		
8		
16		
32		
64		

e. How do the differences between lower and upper approximations for the same number of subintervals compare as the number of subintervals increases?

f. Approximate the distance traveled by an object whose velocity function is $f(x) = 10x^2$ on the interval from 0 to 2.

3. Sketch the graph of $f(x) = \sqrt{4 - x^2}$ for x in $[-2, 2]$.

a. Approximate the area under the graph of $y = f(x)$ for n = 4, 8, 16, 32, and 64 subintervals using lower rectangles and upper rectangles. Complete a table as shown in part (1d).

b. Determine the area under the curve of $y = f(x)$ using a formula from geometry.

c. Does one numeric method of determining the area under the curve provide a better estimate for the area than another numeric method for this function? Explain.

14

Numerical Methods for Approximating Definite Integrals

14.1 Introduction

Finding the antiderivative of a function is often difficult and in many cases impossible. In such cases, numerical methods are necessary to approximate the value of a definite integral for that function. Two frequently used methods in elementary calculus are the midpoint rule and the trapezoid rule. These rules are based on the formulas for the area of a rectangle and the area of a trapezoid, respectively. For many functions, a more accurate numerical approximation for an integral may be obtained using Simpson's Rule. Simpson's Rule is a combination of the midpoint and trapezoid rules.

In this investigation, general formulas are determined for the midpoint, trapezoid, and Simpson's Rule approximations of a definite integral. These are developed through the use of geometric regions sketched on graphs for a small number of subintervals, then generalized to any number n of subintervals.

In applications, it is important to know how large an error occurs when using a numerical method to approximate a definite integral. For that reason, error bounds are also explored.

14.2 The Midpoint and Trapezoid Rules

1. a. The midpoint rule uses areas of rectangles to approximate the area under the graph of $y = f(x)$. Write the formula for the area of a rectangle.

   b. The trapezoid rule uses areas of trapezoids to approximate the area under the graph of a function $y = f(x)$. Write the formula for the area of a trapezoid.

2. Consider the definite integral $\int_1^5 \frac{1}{x}\,dx$. Using the analytical methods for integration from Calculus I, is it possible to use the Fundamental Theorem of Calculus to evaluate this definite integral? Why or why not?

3. Following are two graphs of the function $f(x) = \frac{1}{x}$. The x-interval [1, 5] has been subdivided into four equal-width subintervals. The graph below left shows how four rectangles whose height is determined by the function value of the midpoints of these subintervals can be used to approximate the area under the graph of $y = f(x)$ over the interval [1, 5]. The graph below right shows how four trapezoids determined by these subintervals can be used to approximate this area.

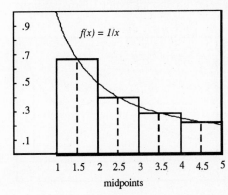
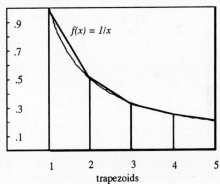

   a. i.   Label the graph on the left with the heights and widths of each rectangle.

       ii. Determine the sum of the areas of these four rectangles.

   b. i.   Label the heights of each of the vertical sides of the trapezoids in the graph above right.

       ii. Label the widths of each trapezoid.

       iii. Determine the sum of the areas of these four trapezoids.

   c. Explain why the values found in parts (3a ii) and (3b iii) give approximations for $\int_1^5 \frac{1}{x}\,dx$.

d.   Which method would provide the most accurate estimate for the area under the curve of $f(x) = \dfrac{1}{x}$ over the interval [1, 5]? Explain.

4.   To help develop the midpoint and trapezoid formulas for approximating definite integrals, the following notation associated with subdivisions of intervals is introduced. An interval [a, b] along the x-axis can be subdivided into n equal-width subintervals using (n + 1) points. The endpoints of each subinterval are designated by $a = x_0 < x_1 < x_2 < \cdots < x_{n-1} < x_n = b$.

The symbol m_i denotes the midpoint of the subinterval $[x_{i-1}, x_i]$. The diagram below shows an interval divided into 4 equal-width subintervals along with the 4 midpoints of the subintervals, m_1, m_2, m_3, and m_4.

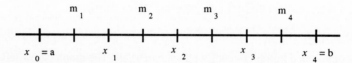

a.   For n equal-width subintervals, what is the width h of each subinterval $[x_{i-1}, x_i]$ in terms of a, b, and n?

b.   Each of the graphs below shows the graph of a function f on the interval [a, b]. The interval is divided into 4 equal-width subintervals.

   i.   On graph (1), draw the rectangles for the midpoint approximations using these 4 subintervals.

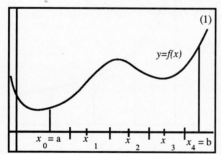
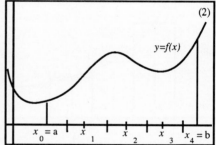

   ii.   Using this subdivision of [a, b], write an expression for M_4, the midpoint approximation using 4 subintervals. Write the formula for the sum of the areas in terms of the width of the subintervals h; and the heights of each rectangle, $f(m_1)$, $f(m_2)$, $f(m_3)$, and $f(m_4)$ where m_i represents the midpoint of the ith subinterval.

   iii.   On graph (2), draw the trapezoids for the trapezoid approximations using these four subintervals.

   iv.   Using this subdivision of [a, b], write an expression for T_4, the trapezoid approximation using 4 subintervals. Hint: The area of the first trapezoid is $\dfrac{h}{2}(f(a) + f(x_1)) = \dfrac{h}{2}(f(x_0) + f(x_1))$. Write the formula for the sum of the areas in terms of h, $f(x_0), f(x_1), f(x_2), f(x_3)$, and $f(x_4)$.

c. i. Divide the interval [a, b] into n equal length subintervals using the points

$$a = x_0 < x_1 < x_2 < \cdots < x_{n-1} < x_n = b.$$

ii. Generalize part (4b ii) to determine an expression for M_n, the midpoint approximation for the definite integral using n subintervals.

iii. Generalize part (4b iv) to determine an expression for T_n, the trapezoid approximation using n subintervals.

14.3 Simpson's Rule

When a definite integral is approximated, the general form is

$$\int_a^b f(x)\, dx = \text{(approximation)} + \text{(error)}.$$

In applications where the value of an integral cannot be determined exactly, it is often important to know how large the error can be. The idea is to determine a maximum value for the absolute value of the error. For n subintervals of [a, b], let ME_n and TE_n represent the errors for the midpoint approximation and trapezoid approximation, respectively. Following are numerical investigations concerning these errors using definite integrals whose exact value can be determined. In these cases, the actual errors introduced by using the midpoint and trapezoid approximations can be investigated.

1. a. Find the exact value of $\int_0^5 x^4\, dx$.

b. Using a numerical approximation technique, this definite integral can be written as

$$\int_0^5 x^4\, dx = \text{(approximation)} + \text{(error)}$$

or equivalently,

$$\text{(error)} = \int_0^5 x^4\, dx - \text{(approximation)}$$

Complete the table on page 82 using this formula with the midpoint and trapezoid approximations.[1] The table entries should be accurate to at least five significant digits.

---

[1] The program INTEGRAL, used to calculate M_n and T_n, can be found in the appropriate appendix. Terminate the program after running it for a given value of n. The values for M_n and T_n are stored in memories M and T, respectively.

n	MEₙ	TEₙ
4		
8		
16		
32		

c. What is happening to the size of the absolute value of the error terms as n increases? Does this seem reasonable? Explain.

d. For each value of n, compare ME_n and TE_n. How are ME_n and TE_n related? Is there any approximate relationship between these two values?

2. Repeat part (1) for the definite integral $\int_0^\pi \sin x \, dx$.

3. For a general definite integral, $I = \int_a^b f(x) \, dx$, numerical methods approximate the value of the integral giving I = (approximation) + (error). For the midpoint and trapezoid rules, respectively, this expression becomes

$$I = M_n + ME_n$$

$$I = T_n + TE_n$$

For the definite integrals in parts (1) and (2), it was observed that $TE_n \approx -2(ME_n)$. Since the two errors are negatives of each other, it seems reasonable that there should be some way of combining the midpoint and trapezoid approximations to reduce the error term, thereby obtaining a better approximation.

a. Write the approximate relationship between the error terms as an equation $TE_n = -2(ME_n)$. Substitute this value into the system of equations above. Solve the system of equations for I.

If TE_n was actually equal to $-2(ME_n)$, this solution for I would completely eliminate the error, and we would have the exact value of I. However, this relationship between the error terms is really an approximation. As a result, the solution for I produces a better approximation for the definite integral than either M_n or T_n.

As the following diagram for 4 subintervals shows, this approximation actually uses 2n subintervals of the interval [a, b], because it uses n subintervals for the trapezoid rule and the n midpoints of these subintervals for the midpoint rule.

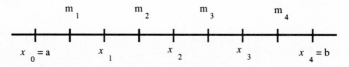

This approximation is known as **Simpson's approximation** with 2n subintervals of equal width. It will be denoted by S_{2n}. In most calculus texts, the method for calculating Simpson's approximation is developed through approximating the graph of a function with parabolas rather than the straight lines used in the midpoint and trapezoid approximations. It makes sense that a better approximation would be obtained by estimating the area under a curve using curved regions rather than polygonal regions. In part (4) below, it will be shown that the method above and the textbook method produce the same result.

b.   Calculate M_4, T_4, and S_8 for $\int_0^5 x^4 \, dx$. Which one is the best approximation? Explain.

c.   Calculate S_{64} for $\int_0^5 x^4 \, dx$. What is the error using this approximation? Explain.

4.   In Section 14.2 part (4c), general formulas were developed for M_n and T_n. It is now possible to develop a formula for S_{2n} to approximate $I = \int_a^b f(x) \, dx$.

a.   Divide the interval [a, b] into 8 equal-width subintervals using the points

$$a = x_0 < x_1 < x_2 < x_3 < x_4 < x_5 < x_6 < x_7 < x_8 = b.$$

Using the points $a = x_0, x_2, x_4, x_6, x_8 = b$, it is possible to obtain a subdivision of the interval [a, b] into 4 equal-length subintervals.

   i.   On the graphs below, draw the rectangles and the trapezoids for the midpoint and trapezoid approximations using the 4 subintervals determined by $a = x_0, x_2, x_4, x_6, x_8 = b$.

   ii.   Label heights and widths of geometric figures on the graphs below.

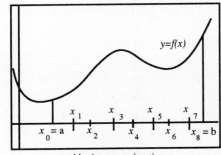
midpoint approximation

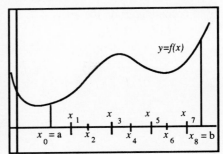
trapezoid approximation

   iii.   Using this subdivision of [a, b], write expressions for M_4 and T_4. The notation used in these expressions can be simplified by letting $y_i = f(x_i)$ for each i and $h = \dfrac{b - a}{8}$.

b. Use part (4a iii) and the rule $S_{2n} = \frac{2}{3} M_n + \frac{1}{3} T_n$ (found in part (3a)) to write an expression for S_8 using this subdivision of [a, b].[2]

c. Divide the interval [a, b] into 2n equal-width subintervals using the points

$$a = x_0 < x_1 < x_2 < \cdots < x_{2n-2} < x_{2n-1} < x_{2n} = b.$$

Let $y_i = f(x_i)$ for each i. Repeat parts (a) and (b) and write formulas for M_n, T_n, and S_{2n}. Use $h = \dfrac{b - a}{2n}$.

The formula you obtain for S_{2n} should be the formula found in most calculus textbooks for Simpson's rule approximation using 2n subintervals of equal width.

14.4 Error Bounds

To determine how accurate an approximation is for

$$\int_a^b f(x)\, dx = (\text{approximation}) + (\text{error}),$$

it is necessary to determine a maximum value for the absolute value of the error. For n subintervals of [a, b], let ME_n, TE_n, and SE_n represent the errors for the midpoint approximation, trapezoid approximation, and Simpson's approximation, respectively. Theorems from advanced calculus provide the following results.

If f'' is continuous on the interval [a, b] and if $|f''(x)| \le K$ for all x in [a, b], then

$$|ME_n| \le \frac{K(b-a)^3}{24n^2}$$

and

$$|TE_n| \le \frac{K(b-a)^3}{12n^2}.$$

If $f^{(4)}$ is continuous on the interval [a, b] and if $|f^{(4)}(x)| \le M$ for all x in [a, b], then

$$|SE_n| \le \frac{M(b-a)^5}{180n^4}.$$

---

[2] This is the formula that is used in the program INTEGRAL to determine Simpson's approximation using 2N subintervals.

Although it is not possible to prove these results with the tools of elementary calculus, it is possible to do some numerical work with actual errors and to observe that these actual errors are consistent with these formulas.

1. a.   For the definite integral $\int_0^5 x^4\, dx$, add a column to the table in Section 14.3 part (1b) for SE_n.[3]

n	ME_n	TE_n	SE_n
4			
8			
16			
32			

b.   What happens to the size of the absolute value of the error terms as n increases? Is this consistent with the formulas for the bounds on $|TE_n|$, $|ME_n|$, and $|SE_n|$? Explain.

c.   For various values of n, divide ME_{2n} into ME_n. By about what whole number factor is $|ME_n|$ divided by when n is doubled? Is this consistent with the formula for $|ME_n|$? Explain.

d.   Repeat part (c) for TE_n.

e.   Repeat part (c) for SE_n.

f.   Repeat parts (a) through (e) for the definite integral $\int_0^\pi \sin x\, dx$.

When using the formulas for the bounds on the absolute value of the errors given in part (1), it is necessary to find upper bounds for $|f''(x)|$ and $|f^{(4)}(x)|$ for x in the interval [a, b]. These are represented by K and M, respectively. In theory, there should be a smallest value for K and M. In practice, these values are often very difficult to find. These are simply estimated by the best value of K and M that can be found. When it is possible to obtain formulas for $y = f''(x)$ and $y = f^{(4)}(x)$, one way of obtaining reasonable values for K and M is to sketch the graphs of $y = f''(x)$ and $y = f^{(4)}(x)$ on the interval [a, b] and determine maxima for the absolute values of the derivatives.

2.   Let $f(x) = \dfrac{1}{x}$.

a.   Use graphs to determine an upper bound for $y = |f''(x)|$ and $y = |f^{(4)}(x)|$ for x in the interval [1, 5].

---

[3]   Use the program INTEGRAL to calculate M_n, T_n, and S_{2n}. Terminate the program after running it for a given value of n. The value for S_{2n} can be accessed by immediately pressing the $\boxed{\text{Ans}}$ key. The values for M_n and T_n will be stored in memories M and T respectively.

b. Estimate $\int_1^5 \dfrac{1}{x} \, dx$ using M_{50}. Determine a bound on $|ME_{50}|$. Between which two values does the value of this definite integral lie?

c. Repeat part (b) using T_{50} and TE_{50}.

d. Repeat part (b) using S_{50} and SE_{50}.

e. Repeat parts (a) through (d) for the definite integral $\int_0^1 \sin(x^2) \, dx$.

14.5 Related Problems

1. a. Sketch $y = \dfrac{1}{x + 4}$ and shade the region being used to estimate the integral $\int_1^3 \dfrac{1}{x + 4} \, dx$.

   b. Using 4 equal-width subintervals on [1,3], estimate the definite integral in part (a) using trapezoids.

2. Use the trapezoid rule with n = 10 to approximate $\int_0^2 \sqrt{1 + x^2} \, dx$.

   a. Write $\int_0^2 \sqrt{1 + x^2} \, dx = T_{10} + TE_{10}$.

   b. Determine a bound on $|TE_{10}|$, the error term in this approximation of the definite integral.

15

Functions Defined by Definite Integrals

15.1 Introduction

The Fundamental Theorem of Calculus states that if f is a function that is continuous on an open interval I containing the number a, then the function F defined by

$$F(x) = \int_a^x f(t) \, dt$$

is differentiable on I, and

$$F'(x) = \frac{d}{dx} \left\{ \int_a^x f(t) \, dt \right\} = f(x).$$

Certain functions can only be defined in terms of definite integrals, rather than a more familiar form not involving a definite integral. For example, the normal probability distribution function (see Section 15.4) that is used commonly in statistics is defined as a definite integral and has no other closed form. Another common function that is defined as a definite integral of a function that has no antiderivative is investigated in Section 15.2. Properties of functions defined by definite integrals are investigated so that you may begin to develop an understanding of such functions and to illustrate what information can be found to determine the behavior of such functions.

15.2 $L(x) = \int_1^x \frac{1}{t}\, dt$

For $x > 0$, define the function L by $L(x) = \int_1^x \frac{1}{t}\, dt$.

1. Sketch the graph of the function $f(t) = \dfrac{1}{t}$ on the x-interval $(0, 10]$.

2. On the graph, indicate the graphic meaning of $y = L(x)$ when $x > 1$.
   a. Is $L(4)$ positive or negative? Explain.
   b. Is $L(0.5)$ positive or negative? Explain.

3. Approximate $L(1.5)$ to the nearest thousandth by using a numerical method for integration. Explain the procedure you used to determine your approximation.

4. Use Simpson's rule with 20 subintervals to complete the following table with numerical approximations for $y = L(x)$.

x	$L(x)$	x	$L(x)$
0.25		3	
0.5		4	
0.75		5	
1		7	
1.5		10	
2		15	

5. Use the values in the table in part (4) to sketch a graph of the function $y = L(x)$ for $x > 0$.

6. a. From the graph, does the function L appear to be increasing or decreasing?
   b. Does the graph of L appear to be concave up or concave down?
   c. Does the graph of L appear to have a maximum or minimum over its domain? Explain.

7. a. On the axes used in part (1), sketch the derivative of L by approximating the slopes of tangents to L (use another color please).
   b. How does your sketch of the derivative of L compare with the graph of f?
   c. Use the Fundamental Theorem of Calculus to determine an expression for $y = L'(x)$. Compare this result with the result of part (7b).
   d. Determine over which intervals the function L is increasing and where it is decreasing. Explain your results.
   e. Determine any maxima and/or minima for L. Explain your results.

8. a. Determine an expression for $y = L''(x)$.

    b. From L'', determine where the function L is concave up and where it is concave down.

    c. Are there any inflection points on the graph of L?

    d. Are these results consistent with the graph in part (5)?

9. a. Determine values for $L(1)$ and $L'(1)$. Explain how you determined these values.

    b. Determine the equation for the tangent line to the graph of L through $(1, L(1))$.

    c. Use a tangent line approximation to estimate $L(1.5)$.

    d. Does this approximation agree with the approximation for $L(1.5)$ shown in the table in part (4)?

    e. Should these approximations agree? Explain.

15.3 $F(x) = \int_0^x \sqrt{1 + t^2}\, dt$

Define $F(x) = \int_0^x \sqrt{1 + t^2}\, dt$.

1. Sketch the graph of $f(t) = \sqrt{1 + t^2}$. Label the x and y scales on the axes.

2. On the graph, indicate the graphic meaning of $y = F(x)$ when $x > 0$.

3. Is $F(-2)$ positive or negative? Explain.

4. Use Simpson's rule with 20 subintervals to complete the following table with numerical approximations for $F(x)$.

x	$F(x)$	x	$F(x)$
10		-10	
5		-5	
4		-4	
3		-3	
2		-2	
1		-1	
0			

5. Use the values in the table in part (4) to sketch a graph of the function F.

6. a. From the graph, determine intervals over which the function F appears to be increasing or decreasing.

    b.   From the graph, determine intervals over which the function F appears to be concave up or concave down.

    c.   Does F appear to have any maxima or minima? Explain.

7.   a.   On the axes in part (5), sketch the derivative of F by approximating the slopes of tangents to F (use another color please).

    b.   How does your sketch of the derivative of F compare with the graph of f?

    c.   Use the Fundamental Theorem of Calculus to determine $y = F'(x)$. Compare this result with the result of part (7b).

    d.   Determine where the function F is increasing and where it is decreasing.

    e.   Determine all maxima and minima for F.

8.   a.   Determine an expression for $y = F''(x)$.

    b.   From F'', determine where the function F is concave up and where it is concave down.

    c.   Are these results consistent with the graph in part (5)?

15.4 The Normal Curve

The function $p(x) = \dfrac{1}{\sqrt{2\pi}}\, e^{-x^2/2}$ is one of the many functions that has no elementary antiderivative. This function is called the standard normal distribution curve. It can be used to model test scores that are *normalized* if the mean (average) and standard deviation (spread) are known.

1.   Sketch the graph of $p(x) = \dfrac{1}{\sqrt{2\pi}}\, e^{-x^2/2}$. Label the graph and the axes.

2.   Complete the table below, then graph $F(x) = \displaystyle\int_{-2}^{x} \dfrac{1}{\sqrt{2\pi}}\, e^{-t^2/2}\, dt$ on the same set of axes as the graph in part (1). Describe the technique you used to determine the table values. Be specific.

x	$F(x)$	x	$F(x)$
10		−10	
5		−5	
4		−4	
3		−3	
2		−2	
1		−1	
0			

3. Determine whether or not $y = F(x)$ is increasing or decreasing. Provide intuitive support and analytical verification for your conclusions.

4. Determine the concavity of F intuitively. Support your observations analytically.

5. a. Find the area under p from -1 to 1. Explain your process.

   b. Find the area under p from -2 to 2. Explain your process.

   c. The area under the curve of p for x in $(-\infty, \infty)$ is 1. Suggest how this area might be estimated.

   d. What percent of the area is accounted for in part (5a)? In part (5b)?

6. Suppose a set of calculus test scores for a particular exam are known to be normal with an average score $\overline{X} = 50$ and a standard deviation $\sigma = 20$. The average score of 50 will be normalized to $\dfrac{X - \overline{X}}{\sigma} = \dfrac{50 - 50}{20} = 0$. A score of 70 is normalized to $\dfrac{70 - 50}{20} = 1$. If the likelihood of obtaining a normalized score of at most x is $P(x)$ where

$$P(x) = \int_{-\infty}^{x} \frac{1}{\sqrt{2\pi}} \, e^{-t^2/2} \, dt$$

   a. How likely is it that a student will receive a score higher than 70 for this exam? (Use an estimate for $P(1)$ to help you find this.) Explain.

   b. How likely is it that a student will receive a score greater than 90? Explain

   c. How likely is it that a student will receive a score less than 35? Explain.

15.5 Related Problems

1. The graph of a function $y = f(t)$ is shown on the right. Define

$$F(x) = \int_{0}^{x} f(t) \, dt \quad \text{for } x \geq 0.$$

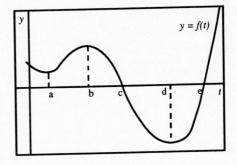

   a. Is $F(c)$ greater than or less than $F(b)$? Explain.

   b. Is $F(d)$ greater than or less than $F(c)$? Explain.

   c. Is $F(e)$ greater than or less than $F(d)$? Explain.

   d. Determine all relative extrema for the function F. Explain your choice(s).

e. Determine all points of inflection for the function F. Explain your choice(s).

f. Draw a rough sketch of the graph of $y = F(x)$.

2. The graph of $y = f(t)$ is given at the right. The function $y = f(t)$ represents the rate of sales of an innovative calculus workbook in its first several years.

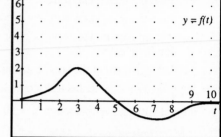

a. What does $\int_0^x f(t)\, dt$ represent in this situation?

b. Estimate the number of workbooks sold during the first five years that the workbook was available. Explain your process and results.

c. From the graph of $y = f(t)$, sketch the graph of $F(x) = \int_0^x f(t)\, dt$.

d. During what year were the most books sold? Explain.

3. As an oven that was preheated to $360°$ heats and cools, its temperature T changes at the **rate R** of $R(x) = \cos(x) \cdot e^{\,\sin(x)}$ with respect to time x.

a. Define the temperature function T as a definite integral.

b. Explain how you could obtain a graph for the temperature function without determining the antiderivative of this function.

c. Determine an expression (without an integral sign) for the function that gives the temperature of the preheated oven at any time x if the hottest temperature is $360°$ ($x = 0$ is chosen to be a time when the oven has been fully heated and is now in the heat/cool cycle).

d. When is the oven hottest? coolest? Explain.

16

Inverse Functions and Relations

16.1 Introduction

For a function modeling a situation, there are situations in which the function value is known and the corresponding x-value is needed. This is particularly true of situations involving trigonometric functions. For example, the height of a bridge is known. The distance of a vehicle from the bridge is known. At what angle from vertical must a camera be aimed from the bridge so that the vehicle is in view? If the vehicle is moving, at what rate must the angle of the camera change to keep the vehicle in view? The first question is a straightforward use of an inverse function. The second requires the use of a derivative of an inverse function.

This investigation allows students to review the concept of inverse relations and functions by first considering the action on points in the coordinate plane. Students then consider the same action on functions. Continuity and differentiability of inverses are considered with respect to the related functions. Also, the relationship to implicit differentiation is made.

16.2 Symmetry from Interchanging *x* and *y*

1. Graph each of the ordered pairs listed on the axes provided. Label these with the letters of the alphabet indicated.

    A (1, 2)   B (3, –2)   C (–1, –1)   D (0, –1.5)

2. Interchange the roles of x and y in the ordered pairs in part (1) so that the ordered pair (a, b) produces the ordered pair (b, a).

3. Graph each of the ordered pairs
    found in part (2) on the axes
    provided. Label these as A', B ',
    C ', and D ' respectively.

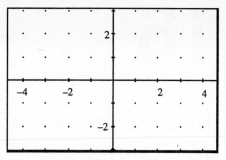

4. Each of the pairs of points A and
    A', B and B ' etc. are reflections of
    each other across the same line.
    What line does this appear to be?
    Determine the equation and sketch
    this line on the axes provided.

5. Plot a point P (a, b) in the first quadrant, a ≠ b. Plot the point P ' (b, a) on the
    same set of axes. Prove that P and P ' are reflections of each other across the
    line found in part (4) as follows.

    a. Determine the midpoint of the line segment PP '. Call it M.

    b. Show that M lies on the line ℓ found in part (4).

    c. Determine the slope of PP '. Show that line segment PP ' is perpendicular
        to line ℓ.

    d. State how the results of parts (5a) through (5c) prove that P ' is the
        reflection of P across line ℓ.

16.3 Inverse Relations and Functions

Let $y = f(x)$ be a function. The set of all ordered pairs of the form (a, b) where
$b = f(a)$ form the graph of the function f. If the coordinates of each of these ordered
pairs are interchanged, the resulting ordered pairs would form a relation called the
inverse of the function f. That is, the inverse of the function f consists of all
ordered pairs of the form (b, a) where $b = f(a)$. The inverse is called the inverse
relation because the set of these ordered pairs is not always a function.

1. For each of the following functions, sketch the graph of $y = f(x)$. Also sketch
    the graph resulting from interchanging the roles of x and y in f, the graph of
    the inverse relation of f.[1]

    a $f(x) = 2x$ b. $f(x) = |2x|$ c. $f(x) = x^2$
    d. $f(x) = x^3$ e. $f(x) = \sin x$ f. $f(x) = \tan x$

2. Use the graphs of the inverses of the functions from part (1) to determine
    which of these inverse relations are functions. Explain your results.

3. What property must the graph of a function have to guarantee that its inverse
    relation gives y as a function of x? Explain.

---

[1] The program INVERSE, provided in the appropriate appendices, interchanges the roles
of x and y and sketches the graph of the inverse relation $(f(x), x)$ for the function
$y = f(x)$ stored in Programs 0 and 1 (CASIO 7000), f_1 (CASIO 7700), or Y_1 (TI).

4. Determine the domain and range of each of the functions in part (1). Determine the domain and range of each corresponding inverse relation. How do the domains and ranges of a function and its inverse compare? Explain.

5. The graph of the inverse relation is the set of points $(f(x), x)$. For the inverse relation to be a function when $y = f(x)$ is a function, what restrictions must be placed on the domain and/or range of $y = f(x)$?

6. Find the definition of an inverse function in your course text. Relate the definition of an inverse function to the work completed in parts (1) through (5).

16.4 Continuity and Differentiability of Inverses

The continuity and differentiability of functions and their inverses are related. Using the definition of continuity and differentiability, we can determine the continuity and differentiability of the inverse function f^{-1} from the properties of the original function $y = f(x)$.

1. a. Graph $f(x) = x^3$. Is it continuous? Explain.

    b. Graph the inverse of $f(x) = x^3$. Is this inverse relation a function? Explain. If not, restrict the domain of f so that the inverse function f^{-1} exists.

    c. Is the inverse function f^{-1} continuous? Explain using the definition of continuity and the relationship between the inverse and the function $y = f(x)$.

2. A function is said to be invertible if its inverse relation is a function. For an invertible continuous function f, is it reasonable to expect that its inverse f^{-1} is continuous? Explain using the definition of continuity.

3. a. Determine the slope of the tangent to the graph of $f(x) = x^3$ at the point $(2, f(2))$.

    b. What point on the inverse is the reflection of the point $(2, f(2))$ on the graph of $f(x) = x^3$?

    c. Explain how the information found in part (3a) can be used to determine the slope of the tangent to f^{-1} at the point found in part (3b).

    d. Determine an equation for the inverse of $f(x) = x^3$ by interchanging the roles of x and y in the expression $y = x^3$ and solving for y.

    e. Determine the derivative of the function found in part (3d) at the point determined in part (3b). How does the value of the derivative compare with your results in part (3c)?

4. a. How would you expect the derivative of a function f to be related to the derivative of its inverse function f^{-1}? Explain.

    b. Would the relationship found in part (4a) hold if the inverse were not a function? Explain any similarities and/or differences.

5. Consider the function $f(x) = x^3 + 3x^2 - 2$ for x in $[-6, 6]$ and y in $[-4, 4]$.

    a. Sketch the graph of the function $y = f(x)$ on the axes provided.

b. Carefully sketch the graph of
    the inverse of f on the same
    coordinate axes. Label both
    graphs appropriately.

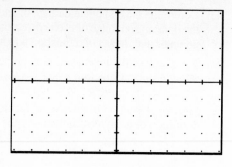

c. Is the inverse of f a function?
    Explain.

d. Determine a *branch* of f over
    which it is invertible. For the
    branch you choose, use the
    largest interval possible for the
    domain of $y = f^{-1}(x)$. Explain.

e. Determine the slope of the tangent at the point $(-2, -3)$ on the graph of
    the inverse of f. Estimate the slope of the tangent from the graph to help
    you decide if your analytic results are correct. Show all of your work,
    including intermediate computer/calculator steps. Explain how you
    obtained your results.

f. Is $y = f^{-1}(x)$ continuous? Explain.

g. Is $y = f^{-1}(x)$ differentiable? In your explanation, include reasons for
    $y = f^{-1}(x)$ being differentiable at certain points and/or not at others.

6. Note that the work completed in parts (4) and (5) provides values of the
    derivatives for inverse relations one point at a time. The formula for the
    derivative of an inverse function can be found for an inverse function whose
    equation can be solved for y. For functions such as the one investigated in part
    (5), implicit differentiation can be used to determine equations for derivatives
    where y is assumed to be a function of x.

    a. Determine an expression for the inverse relation of $y = f(x)$ where
        $f(x) = x^3 + 3x^2 - 2$ by interchanging the roles of x and y.

    b. Find an expression for the derivative of the inverse relation by implicitly
        differentiating the equation found in part (6a). Assume y is a function
        of x.

    c. Determine the slope of the tangent at the point $(-2, -3)$ on the graph of
        the inverse relation. Compare this result with the result from part (5e).

7. In California speeders have been receiving speeding tickets through the mail,
    accompanied by a picture of the car, license plate, and driver. In one location,
    the camera is mounted to an overpass bridge 15 feet above the highway.
    Pictures are taken when the vehicle is at a distance of 100 yards from the base
    of the overpass.

    a. What is the measure of the angle of the camera from vertical at the
        moment the picture is taken?

    b. The car is detected speeding 200 yards before the base of the overpass.

        i.   If a vehicle is traveling at the rate of 75 miles per hour at the time it
             is detected, at what rate must the camera angle change if it is to keep
             the car in view from a distance of 200 yards to a distance of 100 yards?

        ii. At what rate must the camera angle change at the moment the vehicle
             is 150 yards from the base of the bridge?

17

Inverse Trigonometric Functions

17.1 Introduction

This exploration is designed to provide students with more experience in working with inverse trigonometric functions than is usually provided in calculus texts. In this exploration, best choices of principal branches are determined for the trigonometric functions so that their inverses are functions; the relationships among the inverse trigonometric functions are investigated, and relationships between inverse trigonometric functions and the trigonometric functions are explored.

17.2 Choosing Principal Branches

The selection of the principal branch of a trigonometric function in order to define an inverse function is somewhat arbitrary. This is especially true for the inverse secant and inverse cosecant functions.

1. a. To define the inverse sine function, the interval $[\frac{-\pi}{2}, \frac{\pi}{2}]$ is used as the principal branch of the sine function. Can this interval be used as the principal branch for the cosine function in order to define the inverse cosine function? Explain why or why not.

   b. Why isn't the interval $[0, \frac{\pi}{2}]$ used as the principal branch for the cosine function in order to define the inverse cosine function?

   c. What would be a better choice for a principal branch? Explain.

2. a. Show that the function $f(x) = \sin x$ for x in $[\frac{\pi}{2}, \frac{3\pi}{2}]$ has an inverse that is a function.

   b. In defining the inverse sine function, why it would not be desirable to use the interval $[\frac{\pi}{2}, \frac{3\pi}{2}]$ as the principal branch of the sine function? Explain.

3. a. Use the inverse relationship between the sine and the inverse sine functions, implicit differentiation, and a trigonometric identity to determine the derivative of $f(x) = \text{Sin}^{-1}x$ for the principal branch of the sine function shown in part (1a).

   b. With respect to the derivative, where is $f(x) = \text{Sin}^{-1}x$ increasing? Decreasing? Maximum? Minimum? Show your work and explain your results in terms of the graph of f.

   c. What is the derivative of $f(x) = \text{Sin}^{-1}x$ for the principal branch of the sine function in part (2a)?

   d. Answer part (3b) using the principal branch and derivative in part (3c).

   e. Is there a preference for a principal branch of the sine function based on your work with the derivative of $f(x) = \text{Sin}^{-1}x$? Explain.

4. There are several choices for the principal branch of the function $f(x) = \sec x$. Two common choices provided by calculus textbook authors are

$$D_1:\ [0, \frac{\pi}{2})\ \cup\ (\frac{\pi}{2}, \pi]$$

and

$$D_2:\ [0, \frac{\pi}{2})\ \cup\ [\pi, \frac{3\pi}{2})$$

   a. Find $\frac{d}{dx}(\text{Sec}^{-1}x)$ using the inverse relationship with the secant function, implicit differentiation, and a trigonometric identity.

   b. Determine an expression for $\frac{d}{dx}(\text{Sec}^{-1}x)$ when the principal branch for $y = \sec x$ is chosen to be D_1.

   c. What are the advantages of choosing D_1 as the principal branch for $f(x) = \sec x$? Explain.

   d. Determine an expression for $\frac{d}{dx}(\text{Sec}^{-1}x)$ when the principal branch for $y = \sec x$ is chosen to be D_2.

   e. What are the advantages of choosing D_2 as the principal branch for $f(x) = \sec x$? Explain.

5. In defining the inverse function of a trigonometric function, what properties should the principal branch of the trigonometric function possess? Explain.

17.3 Relationships Among Inverse Trigonometric Functions

1. The functions $y = \text{Csc}^{-1}x$ and $y = \text{Sec}^{-1}x$ are not in the library of functions on most calculators or computer packages.

    a. Obtain accurate hand-drawn graphs of each of these functions from graphs of their respective inverses (use the principal branches for these functions listed in the course text). Explain how you obtained each graph and include well-labeled graphs of all four functions.

    b. Is $\text{Sec}^{-1}x = \dfrac{1}{\text{Cos}^{-1}x}$? Explain.

    c. Sketch $f(x) = \text{Csc}^{-1}x$ and $g(x) = \text{Sec}^{-1}x$ on the same set of axes. Notice that the shapes of these graphs are similar. How can you obtain the graph of g from the graph of f?

    d. From your work in part (1c), suggest a way to write $f(x) = \text{Csc}^{-1}x$ in terms of $g(x) = \text{Sec}^{-1}x$. Explain.

2. For $x > 0$, it is possible to rewrite $y = \text{Csc}^{-1}x$ in terms of the inverse sine function and to rewrite $y = \text{Sec}^{-1}x$ in terms of the inverse cosine function. Complete the following to rewrite $y = \text{Csc}^{-1}x$ in terms of $y = \text{Sin}^{-1}x$.

    a. Let $\theta = \text{Csc}^{-1}x$ for $x > 0$. Sketch a right triangle. Label one of the acute angles as θ. Label the triangle so that $\csc\theta = x$. What is $\sin\theta$ for this triangle?

    b. Using the triangle from part (2a), for $x > 0$, derive a formula for $y = \text{Csc}^{-1}x$ in terms of the inverse sine function. Explain your derivation.

3. a. Use the formulas from part (2) to sketch the graph of $y = \text{Csc}^{-1}x$ for $x > 0$ using a computer or graphing calculator. Does this graph agree with your graph in part (1a)? Explain similarities and/or differences.

    b. Use the formula from part (2) to sketch the graph of $y = \text{Csc}^{-1}x$ using the viewing rectangle $[-5, 5]$ by $[-2, 5]$. Does this graph agree with your graph in part (1a)? Explain similarities and/or differences.

17.4 Cofunction Relationships

The pairs of functions cosine and sine, tangent and cotangent, and secant and cosecant are called **cofunctions** because they satisfy the following cofunction identities: $\sin\theta = \cos(\dfrac{\pi}{2} - \theta)$, $\tan\theta = \cot(\dfrac{\pi}{2} - \theta)$, and $\sec\theta = \csc(\dfrac{\pi}{2} - \theta)$.

Are there similar relationships for their respective inverse functions? This activity will assist in the discovery of one such relationship.

1. Consider the triangle to the right.
    a. Label the triangle so that $x = \tan \theta$.
    b. The cotangent of which angle is x?
    c. Determine an expression for $\theta = \text{Tan}^{-1}x$ with respect to the triangle.
    d. Determine an expression for $\theta = \text{Cot}^{-1}x$ with respect to the triangle.

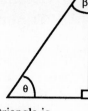

2. Write β in terms of θ. (Recall that the sum of the angles of a triangle is $180° = \pi$ radians.)

3. Write $\text{Cot}^{-1}x$ in terms of θ and then in terms of $\text{Tan}^{-1}x$.

4. The argument in parts (1) through (3) allow the function $y = \text{Cot}^{-1}x$ to be written in terms of $y = \text{Tan}^{-1}x$ for positive values of x. To extend this result to negative values for x, again let $\theta = \text{Tan}^{-1}x$. Then $x = \tan \theta$ for θ in $(-\pi/2, \pi/2)$.
    a. Use the proper cofunction identity to rewrite the equation $\tan \theta = x$ in terms of the $\cot (\pi/2 - \theta)$.
    b. For θ in $[0, \dfrac{\pi}{2})$, between which two angles does $(\pi/2 - \theta)$ lie?
    c. Explain why the inverse cotangent function can be applied to the result in part (4a). Rewrite the equation from part (4a) in terms of the inverse cotangent function.
    d. Write the equation for $y = \text{Cot}^{-1}x$ in part (4c) in terms of $y = \text{Tan}^{-1}x$.

5. Use the result in part (4d) and a computer or graphing calculator to draw a graph of $y = \text{Cot}^{-1}x$. Use a viewing rectangle that shows all the important features of the graph. Label the graph and the axes showing the x and y scales.

6. Use an argument similar to that outlined in parts (1) through (4) to derive an expression that gives $y = \text{Cos}^{-1}x$ in terms of $y = \text{Sin}^{-1}x$.

17.5 Related Problems

From Calculus I, we know that $\lim\limits_{x \to 0} \dfrac{\sin(x)}{x} = 1$.

1. What does $\lim\limits_{x \to 0} \dfrac{\text{Sin}^{-1}(x)}{x}$ seem to be? Explain your process. Show your work including intermediate results.

2. Determine $\lim\limits_{x \to 0} \dfrac{\text{Sin}^{-1}(x)}{x}$ using the inverse function relationship between $y = \sin(x)$ and $y = \text{Sin}^{-1}(x)$. Explain your work.

18

Hyperbolic Functions

18.1 Introduction

The hyperbolic trigonometric functions are linear combinations of the functions $y = e^x$ and $y = e^{-x}$. In fact, the sum of the hyperbolic sine with the hyperbolic cosine is the function $y = e^x$. The hyperbolic trigonometric functions are interesting in that they have similar properties to the trigonometric functions. They are also related to the hyperbola.

This exploration is designed to highlight the symmetry of functions in general and the symmetry of the hyperbolic trigonometric functions in particular. It is also designed to explore the relationships between the hyperbolic trigonometric functions, the trigonometric functions, and the hyperbola.

18.2 Symmetry of Functions

1. An interval along the x-axis that is symmetric about $x = 0$ is either the entire x-axis or an interval of the form $(-a, a)$ where a is a real number. If a function f is defined on such an interval, then the value $f(x)$ is defined if and only if $f(-x)$ is defined.

    Show that every function f that is defined on a symmetric interval about $x = 0$ can be written as

    $$f(x) = \frac{f(x) + f(-x)}{2} + \frac{f(x) - f(-x)}{2}.$$

2. Recall that an even function is a function g such that $g(-x) = g(x)$ for all x in its domain. The graph of an **even function** is symmetric about the y-axis.

Also, an **odd function** is a function h such that $h(-x) = -h(x)$ for all x in its domain. The graph of an odd function is symmetric about the origin.

a. Each of the functions in parts (i) through (iii) below are defined for all x. Determine the symmetry of each of these functions.

    i. $f(x) = x^2$ ii. $f(x) = (x - 1)^3$ iii. $f(x) = \sin x$

b. For each of the following types of symmetry, choose a function whose domain is symmetric about $x = 0$.

    i. even   ii. odd   iii. neither even nor odd

3. Let the function g be defined to be the first component of f in part (1):

$$g(x) = \frac{f(x) + f(-x)}{2}.$$

a. For each of the 6 functions in part (2), write an expression for the function g and sketch the graph of g.

b. Does the graph of g appear to have any symmetry? Explain.

c. How does the symmetry of f compare with that of the function g?

4. Let the function h be defined to be the second component of f in part (1):

$$h(x) = \frac{f(x) - f(-x)}{2}$$

a. For each of the 6 functions f in part (2), write an expression for the function h and sketch the graph of h.

b. Does the graph of h appear to have any symmetry? Explain.

c. How does the symmetry of f compare with that of the function h?

5. a. TRUE or FALSE: Any function f which is defined for all real numbers x can be written as the sum of two functional components, one of which is an odd function, and the other an even function.

    b. If this statement is true, use the definition of even or odd functions to explain why the functional components are even or odd. If the statement is false, give an example of a function f for which it is false.

18.3 Hyperbolic Functions

The functions $f(x) = \sinh x$ and $g(x) = \cosh x$ are called **hyperbolic trigonometric** functions. The following questions are designed to assist students in understanding this designation by relating these functions to the trigonometric functions and to the hyperbola.

1. a. Write $f(x) = e^x$ as the sum of an even function and an odd function, as suggested in section 18.2.

    b. What is another name for the even component of $f(x) = e^x$?

    c. What is another name for the odd component of $f(x) = e^x$?

2. Compare the symmetries of the sine function with the hyperbolic sine function, and the cosine function with the hyperbolic cosine function.

3. To begin to understand the designation *trigonometric* in the names of the hyperbolic trigonometric functions, compare the sine function with the hyperbolic sine function, and the cosine function with the hyperbolic cosine function as follows.

   a. Find the first derivatives of both $f(x) = \sinh x$ and $f(x) = \cosh x$.

   b. Find the second derivatives of both $f(x) = \sinh x$ and $f(x) = \cosh x$.

   c. How do the first and second derivatives of these functions compare to each other? Note any similarities and/or differences.

   d. Prove that $\cosh^2 t - \sinh^2 t = 1$.

   e. How is the relationship in part (3d) similar to that of the trigonometric functions sine and cosine?

4. To begin to understand the designation *hyperbolic* in the names of the hyperbolic trigonometric functions, complete the following.

   a. In part (3d), let $x = \cosh t$ and $y = \sinh t$.

      i. Write the identity in part (3d) as an equation relating the variables x and y.

      ii. Does this equation define y as a function of x? Does this equation define x as a function of y? Explain.

   b. Sketch a graph of the equation found in part (4a i). The graph of this equation is called the **unit hyperbola**. If the points $(\cosh t, \sinh t)$ were to be graphed for all real numbers t, would the unit hyperbola be sketched? Why or why not?

   c. Investigate the relationship between the points $(\cos t, \sin t)$ on the unit circle, and the points $(\cosh t, \sinh t)$ on the unit hyperbola as follows.

      i. Graph the unit circle and the unit hyperbola on the same set of axes.

      ii. Determine the value(s) of t for which $\cos t = \cosh t$ and $\sin t = \sinh t$. Explain how you determined your result.

18.4 Related Problems

1. Sketch the graph of $f(x) = \cosh x$. Be sure to label the scales of the axes.

   a. What is the domain and range of $f(x) = \cosh x$?

   b. Is the inverse of $f(x) = \cosh x$ a function? Explain.

   c. If the inverse of f is not a function, how can f be redefined so that its inverse is a function? Explain.

   d. Sketch the graph of the **function** $f^{-1}(x) = \cosh^{-1}x$ on the axes in part (1).

   e. Over which intervals is $f(x) = \cosh x$ increasing or decreasing? Prove your observation analytically.

    f.   For the principal branch of $f^{-1}(x) = \cosh^{-1}x$ determined in part (1c) is f^{-1} increasing or decreasing? Explain by using the inverse relationship between f and f^{-1}.

2. a.   Identify the domain and range of $f(x) = \operatorname{sech} x$. Verify your observations with an analytical argument using limits and/or derivatives.

    b.   Determine f'. Use f' to determine intervals over which f is increasing or decreasing.

    c.   Determine f''. Discuss the concavity of f.

    d.   From parts (2b) and (2c), determine if f is invertible. If not, choose a principal branch for f so that f^{-1} exists.

19

Indeterminate Forms and L'Hopital's Rule

19.1 Introduction

Indeterminate forms are undefined expressions whose behavior cannot be determined simply by evaluating the expression at the point for which it is undefined or as $|x|$ grows large. Expressions such as $f(x) = \dfrac{\sin x}{x}$ at $x = 0$, $g(x) = \dfrac{x^2 - 1}{x - 1}$ at $x = 1$, and $h(x) = \left(1 + \dfrac{1}{x}\right)^x$ as x tends to infinity are examples of indeterminate forms. More thorough analysis must be completed to determine the behavior of such functions near the point of interest or as $|x|$ tends to infinity. Such functions appeared in Chapter 4, *Behavior at a Point and End Behavior*.

This chapter investigates graphical, numerical, and analytical approaches to determining the behavior of functions that produce the indeterminate forms $\dfrac{0}{0}$ and $\dfrac{\infty}{\infty}$ at a point or as $|x|$ grows large. An understanding of L'Hopital's Rule is developed, particularly where two functions intersect the x-axis.

19.2 Functions Yielding $\dfrac{0}{0}$ Forms

1. Recall your work from Chapter 4 with behaviors of functions. In determining the behavior of $f(x) = \dfrac{\sin x}{x}$ at $x = 0$, the graph of f near $x = 0$ was sketched then a table of values for x near zero was created. To obtain more information, complete the following.

   a. Graph $y = \sin x$ and $y = x$ on the viewing rectangle $[-1.5, 1.5]$ by $[-1, 1]$. As $|x|$ gets very small, how does the graph of $y = x$ compare to that of $y = \sin x$? Describe the appearance of the graphs as compared to each other on a very small interval.

   b. How might the tangent lines to the graphs of $y = \sin x$ and $y = x$ be helpful in determining the quotient $f(x) = \dfrac{\sin x}{x}$ for x near 0?

   c. Create a table for the function $f(x) = \sin x$ as follows. Notice that x decreases by half each time. [1]

x	$f(x)$	x	$f(x)$
0.5	0.4794255386	-0.5	-0.4794255386
0.25		-0.25	
0.125		-0.125	
0.0625		-0.0625	
0.03125		-0.03125	
0.015625		-0.015625	
0.0078125		-0.0078125	

   d. From the table in part (1c), as x approaches zero how do the values of x and $\sin x$ compare?

   e. From the table in part (1c) and the graphs sketched in part (1a), what would you expect the values of $\dfrac{\sin x}{x}$ and $\dfrac{x}{\sin x}$ to be for x close to zero? Why?

   f. Graph $f(x) = \dfrac{\sin x}{x}$. Does the graph of f make sense, considering your response to part (1e)? Explain.

---

[1] Use the program BEHAVIOR in the appropriate appendix to create the table.

2. Consider the function $h(x) = \dfrac{\sin \frac{x}{3}}{e^{2x} - 1}$. It would be interesting to determine how this function behaves near $x = 0$ where it is undefined.

a. Determine $\lim\limits_{x \to 0} \sin \frac{x}{3}$.

b. Determine $\lim\limits_{x \to 0} (e^{2x} - 1)$.

c. Is it possible to evaluate the limit, $\lim\limits_{x \to 0} \dfrac{\sin \frac{x}{3}}{e^{2x} - 1}$ using the theorem for a limit of a quotient? Explain.

3. Graphically investigate the behavior of $y = h(x)$ near $x = 0$ as follows.

a. Sketch a graph of the function $h(x) = \dfrac{\sin \frac{x}{3}}{e^{2x} - 1}$ using the viewing rectangle $[-2.5, 2.5]$ by $[-2, 2]$.

b. Zoom in on the graph of h a few times at the y-intercept of the graph.

c. How does h behave on small intervals containing $x = 0$?

d. What information can be obtained about $\lim\limits_{x \to 0} h(x)$ from this graph?

4. Numerically investigate the behavior of h near $x = 0$ as follows.

a. Complete a table for $y = h(x)$ as in part (1a).

b. Continue the table far enough to determine the behavior of $y = h(x)$ numerically as x approaches zero from both the right and the left. How does $y = h(x)$ behave for x near zero?

c. What information is obtained about the $\lim\limits_{x \to 0} h(x)$ from the table?

5. The investigations in parts (3) and (4) provide some information about the behavior of the function $y = h(x)$ for x near zero. These methods can only be used to make a conjecture about the exact value of $\lim\limits_{x \to 0} h(x)$. The following investigation will help you develop an understanding of a useful and efficient method of determining $\lim\limits_{x \to 0} h(x)$ for such functions as in parts (1) and (2).

For $h(x) = \dfrac{\sin \frac{x}{3}}{e^{2x} - 1} = \dfrac{f(x)}{g(x)}$, consider the behaviors of the functions f and g individually on a small interval containing $x = 0$. Let $f(x) = \sin \frac{x}{3}$ and $g(x) = e^{2x} - 1$.

a. Sketch the graphs of $f(x) = \sin \frac{x}{3}$ and $g(x) = e^{2x} - 1$ both on the viewing rectangle $[-1.5, 1.5]$ by $[-1, 1]$.[2]

---

[2] It is important that the x and y scales be chosen so that tick marks are the same distance apart and represent the same units in both directions. For rectangular screens

b. Sketch f and g again on $[-0.15, 0.15]$ by $[-0.1, 0.1]$. Describe the appearance of the graphs of f and g on this viewing rectangle.

c. On this small viewing rectangle, how would the graphs of the tangent lines to f and g through the origin compare with the graphs of f and g respectively?

d. As x approaches zero, how do f and g appear to be changing with respect to each other?

6. a. Estimate the slopes of the functions f and g at $x = 0$ on a viewing rectangle $[-0.15, 0.15]$ by $[-0.1, 0.1]$. Divide your estimate of the slope of f by your estimate of the slope of g. How does this quotient compare with your answer to part (5d)?

b. Might the slopes of tangent lines to f and g through $x = 0$ be helpful in determining $\lim_{x \to 0} h(x)$? Explain.

7. a. Determine the equation for the line tangent to the graph of $f(x) = \sin \frac{x}{3}$ through the point $(0, 0)$. Write the equation for this tangent line as a function $y = T_f(x)$.

b. Repeat part (7a) for the function $g(x) = e^{2x} - 1$, writing the tangent line through $(0, 0)$ as the function $y = T_g(x)$.

c. What do the slopes of these two lines represent?

d. Determine the value of the limit $\lim_{x \to 0} \dfrac{T_f(x)}{T_g(x)}$.

8. Compare the value of the limit in part (7d) with the value of $\lim_{x \to 0} \dfrac{\sin \frac{x}{3}}{e^{2x} - 1}$ from parts (3d) and (4c). Are these the same? Should they be? Explain.

9. To determine the limit $\lim_{x \to a} \dfrac{f(x)}{g(x)}$ where $\lim_{x \to a} f(x)$ and $\lim_{x \to a} g(x)$ are both zero, is $\lim_{x \to a} \dfrac{f'(x)}{g'(x)}$ helpful? Explain in light of your responses to parts (5) through (8).

19.3 $\frac{0}{0}$ and $\frac{\infty}{\infty}$ Forms

1. a. Consider the functions $f(x) = \ln(x + 1)^2$ and $g(x) = x$. Graph $f(x) = \ln(x + 1)^2$ and $g(x) = x$ on a viewing rectangle $[-5, 5]$ by $[-3, 3]$.

---

where $x \sim 1.5y$, use the dimensions given. If the graphics screen is square, let the x and y dimensions be equal.

b. Zoom in[3] on f and g at $x = 0$ until both f and g appear as straight lines. Sketch this view on the graphs. Label the functions and the axes.

c. From the second graph, estimate $\lim\limits_{x \to 0} \dfrac{f(x)}{g(x)}$ as in Section 19.2 part (6). Explain your conclusion.

2. Determine $\lim\limits_{x \to 0} \dfrac{f(x)}{g(x)}$ analytically as in Section 19.2 part (7). Show your work.

3. What is the relationship between the view of the functions f and g in the second graph and the analytical work shown in part (2)? Explain.

4. Sketch the graphs of f and g on a viewing rectangle [–50, 50] by [–30, 30].

5. What do you expect $\lim\limits_{x \to \infty} \dfrac{f(x)}{g(x)}$ to be? Explain in terms of the individual behaviors of f and g as x grows large.

6. As x grows large, might the rates of change of the functions f and g be useful in determining $\lim\limits_{x \to \infty} \dfrac{f(x)}{g(x)}$? Explain.

7. Find $\lim\limits_{x \to \infty} \dfrac{f'(x)}{g'(x)}$. How does this quantity compare with your results from part (5)?

8. Look up L'Hopital's Rule in your course text. Explain the hypotheses and results of L'Hopital's Rule from the results of your work in Sections 19.2 and 19.3.

19.4 Related Problems

1. Suppose $f(c) = a = g(c)$ as shown in the graph to the right. Is L'Hopital's Rule useful in determining $\lim\limits_{x \to c} \dfrac{f(x)}{g(x)}$? Explain.

    (Do not restate the theorem. Note that the theorem does not apply when $\dfrac{f(x)}{g(x)}$ is not in $\dfrac{0}{0}$ or $\dfrac{\infty}{\infty}$ form. Reason out why L'Hopital's Rule is or isn't applicable here.)

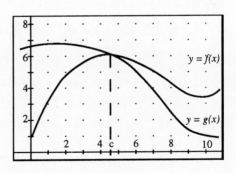

---

[3] While zooming in on the graphs, take care to change both the x and y axes by the same factor so as not to distort the slopes of the graphs.

2. The graphs of the functions f and g are sketched on the viewing rectangle [–5, 5] by [–3, 3]. For the questions that follow, briefly justify each conclusion or state why a conclusion cannot be made.

    a. Sketch the graph of $h(x) = \dfrac{f(x)}{g(x)}$ for the graphs of f and g provided in the figure below.

    b. Using the information presented in this graph, what conclusions can you make about each of the following limits?

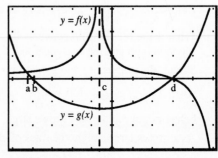

    i. $\displaystyle\lim_{x\to a} \frac{f(x)}{g(x)}$

    ii. $\displaystyle\lim_{x\to b} \frac{f(x)}{g(x)}$

    iii. $\displaystyle\lim_{x\to c} \frac{f(x)}{g(x)}$

    iv. $\displaystyle\lim_{x\to d} \frac{f(x)}{g(x)}$ v. $\displaystyle\lim_{x\to\infty} \frac{f(x)}{g(x)}$ vi. $\displaystyle\lim_{x\to\infty} \frac{f(x)}{g(x)}$

    c. Sketch the graph of

    $h(x) = \dfrac{g(x)}{f(x)}$ for the graphs of f and g provided in the figure.

    d. Using the information presented in this graph, what conclusions can you make about each of the following limits?

    i. $\displaystyle\lim_{x\to b} \frac{g(x)}{f(x)}$ ii. $\displaystyle\lim_{x\to c} \frac{g(x)}{f(x)}$ iii. $\displaystyle\lim_{x\to d} \frac{g(x)}{f(x)}$

3. Use the following definition to answer the questions below.

Definition: Rates of Growth as $x\to\infty$

    I.   If $\displaystyle\lim_{x\to\infty} f(x) = \pm\infty$ and $\displaystyle\lim_{x\to\infty} g(x) = \pm\infty$, then f **grows faster than** g as

    $x\to\infty$ if $\displaystyle\lim_{x\to\infty} \frac{f(x)}{g(x)} = \infty$ or equivalently, if $\displaystyle\lim_{x\to\infty} \frac{g(x)}{f(x)} = 0$.

    II. If f grows faster than g as $x\to\infty$, we also say that g **grows slower than** f as $x\to\infty$.

    III. f and g **grow at the same rate** as $x\to\infty$ if $\displaystyle\lim_{x\to\infty} \frac{f(x)}{g(x)} = L \neq 0$. (L is finite and not zero.)

    a. Suggest approaches that you might use to begin to determine the comparative rates of growth of two functions. **Be specific.**

    i.   numerically   ii.   graphically   iii.   analytically

    b. For the following problems, first use a graphical or numerical approach to estimate which of the functions f or g grows faster. Then, prove your conjecture analytically. Show all of your work, **including intermediate computer or calculator results.**

i.   Which function grows faster as $x\to\infty$, $f(x) = e^x$, or any polynomial $g(x) = a_nx^n + a_{n-1}x^{n-1} + \dots + a_1x + a_0$?

ii. Which function grows faster as $x\to\infty$, $f(x) = \ln x$, or $g(x) = x^{1/n}$ for any positive integer n?

20

Infinite Series and Improper Integrals

20.1 Introduction

In this exploration, the relationship between an infinite series and an improper integral of a function f from a number a to ∞ will be investigated. An infinite series is a sum of the form $\sum_{i=1}^{+\infty} a_i$. It is not possible to add infinitely many numbers by normal arithmetic operations. Instead, the sum of this series is defined as a limit. In particular, the series is said to **converge** to the sum S if

$$\lim_{n \to +\infty} \sum_{i=1}^{n} a_i = S$$

If this limit does not exist, the series is said to **diverge**.

The finite sum $\sum_{i=1}^{n} a_i$ is called a partial sum of the infinite series.

For example, for the infinite series $\sum_{i=1}^{+\infty} \frac{1}{i^2}$, the first four partial sums are

$$S_1 = 1 \qquad\qquad S_2 = 1 + \frac{1}{4} = \frac{5}{4} = 1.25$$

$$S_3 = 1 + \frac{1}{4} + \frac{1}{9} = 1.36111111 \qquad S_4 = 1 + \frac{1}{4} + \frac{1}{9} + \frac{1}{16} = 1.42361111$$

The sum of a series is the limit of its sequence of partial sums.

20.2 Infinite Series Versus Improper Integrals

1. a. Consider the sequence $\{a_n\} = \{\frac{1}{n}\}$. What are the first 10 terms of this sequence?

   b. Determine the first 5 terms of the sequence $\{S_n\}$ where $S_n = \sum_{i=1}^{n} \frac{1}{i}$.

   c. Determine S_{10}, the tenth term of this sequence.[1]

2. Recall that $\int_a^b f(x)\, dx = \lim_{n \to \infty} \sum_{i=1}^{n} f(t_i)\, \Delta x$ where the interval $[a, b]$ is

   partitioned into n equal-width subintervals with partition points

$$a = x_0, x_1, x_2, \ldots , x_{i-1}, x_i, \ldots , x_n = b;$$

   t_i is an arbitrary element of the interval $[x_{i-1}, x_i]$; and $\Delta x = \dfrac{b-a}{n}$.

   a. Suppose t_i is chosen to be equal to the integer i, and Δx is chosen to be 1.
      Rewrite $\lim_{n \to \infty} \sum_{i=1}^{n} f(t_i)\, \Delta x$ for this case.

   b. Explain how a rough estimate for the definite integral $\int_a^b f(x)\, dx$ might be obtained using the choices of t_i and Δx from part (2a).

   c. Suggest how the area under a curve of infinite length might be estimated by the sum found in part (2a).

3. Estimate the area under the curve of $f(x) = \dfrac{1}{x}$ on the interval $[1, 5]$ as follows.

   a. Sketch $f(x) = \dfrac{1}{x}$ on the axes provided.

   b. Partition the interval $[1, 5]$ using 4 equal-width subintervals.

   c. Sketch rectangles whose heights are determined using the right endpoints of the 4 subintervals. (Note that this

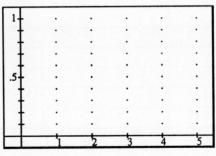

---

[1] As n gets large, the process of determining partial sums is tedious to carry out on a calculator. Since this type of calculation will be done quite frequently, it is advisable to use a program on a computer or calculator that will calculate partial sums of a series. CASIO and TI graphics calculator users should use the program SERIES.

gives an underestimate of the actual area.)

d.   Label the rectangles showing the area of each.

e.   Write the sum of the areas as in Section 20.1. Find the sum of these areas.

f.   Write the sum using the summation notation $\sum_{i=I}^{N} \frac{1}{i}$, and the appropriate

replacements for I and N.

4.   Repeat part (3) using left endpoints to obtain an overestimate of the area under

the graph of $f(x) = \frac{1}{x}$ on [1, 5].

5.   Build a chart showing underestimates and overestimates for the area under

$f(x) = \frac{1}{x}$ as follows. Write the partial sums of the series $\sum_{i=1}^{+\infty} \frac{1}{i}$ that denote

the estimates of the areas of the indicated regions using summation notation.[2]

Evaluate $\int_{1}^{n} \frac{1}{x} \, dx$ using the Fundamental Theorem of Calculus. Record

values accurate to eight decimal digits.

n	underestimate	$\int_{1}^{n} \frac{1}{x} dx$	overestimate
2	$\sum_{i=2}^{2} \frac{1}{i} = \frac{1}{2} = 0.5$	$\int_{1}^{2} \frac{1}{x} dx =$ 0.69314718	$\sum_{i=1}^{1} \frac{1}{i} = 1 = 1$
3	$\sum_{i=2}^{3} \frac{1}{i} = \frac{1}{2} + \frac{1}{3} =$ 0.83333333	$\int_{1}^{3} \frac{1}{x} dx =$ 1.09861229	$\sum_{i=1}^{2} \frac{1}{i} = 1 + \frac{1}{2} = 1.5$
4			
5			
6			
10			
100			
1000			

6.   a.   Compare the sizes of the underestimate, the overestimate, and the actual

area under $f(x) = \frac{1}{x}$ on [1, n].

---

[2] CASIO and TI graphics calculator users should use the program SERIES.

b.   How much does each grow as n changes from 3 to 4? from 6 to 10? from 10 to 100? from 100 to 1000?

c.   Does one seem to grow faster than the other?

d.   How would they grow with respect to each other as n tends to infinity? Explain.

e.   Does the area under $f(x) = \dfrac{1}{x}$ appear to be finite or infinite? Explain.

7.   Complete parts (3) through (6) for $f(x) = x^{-2}$.

8.   Complete parts (3) through (6) for $f(x) = xe^{-x^2}$.

9.   a.   Which of the improper integrals $\displaystyle\int_1^{+\infty} f(x)\, dx$ for the functions in parts (6) through (8) converge? Prove this analytically.

b.   Could the convergence or divergence of the integral $\displaystyle\int_1^{+\infty} f(x)\, dx$ be helpful in determining whether the sum $\displaystyle\sum_{i=1}^{+\infty} f(i)$ is finite or infinite?

i.   If the integral converges, would you expect the series to converge or diverge? Explain.

ii.   If the integral diverges, would you expect the series to converge or diverge? Explain.

20.3 Related Problems

1.   Use a graphical or numerical argument to suggest whether or not $\displaystyle\int_0^{\infty} e^{-x}\, dx$ converges or diverges. Support your findings analytically.

2.   Use the Fundamental Theorem of Calculus to explain why the following is incorrect.

$$\int_0^1 \frac{1}{x}\, dx \;=\; \ln x \,\Big|_0^1 \;=\; \ln 1 - \ln 0 = 1 - (-\infty) = \infty$$

3.   Determine whether or not $\displaystyle\int_0^{\infty} \frac{1}{1 + e^x}\, dx$ converges or diverges. Explain your results. (Hint: Use the results from part (1).)

21

Taylor Polynomials

21.1 Introduction

While exploring linear approximations of functions, we observed that the tangent line to the graph of a function f at the point $(a, f(a))$ closely approximates the graph of the function for small intervals containing $x = a$. (See the figure below.)

The equation for the tangent line T written in function form is

$$T(x) = f(a) + f'(a)(x - a)$$

This equation is also called the **Taylor polynomial of degree 1** for the function f about $x = a$, and is denoted by P_1.

It is reasonable to assume that the approximation of $y = f(x)$ for x near a

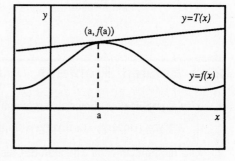

can be improved by considering higher order derivatives. For example, a second degree polynomial P_2, which uses a second derivative of f to take the concavity of f into consideration, might better approximate f than a first degree polynomial $T(x) = P_1(x)$ which only considers the slope of f near $x = a$ and the function value of f at $x = a$. This is exactly the case.

In this exploration, students will determine expressions for the second and third Taylor Polynomials to approximate a function $y = f(x)$ for x near a. These expressions will be generalized to an n-degree Taylor Polynomial. Taylor Polynomials will then be used to approximate the functions $f(x) = e^x$ and $f(x) = \sin x$ for x near zero.

21.2 Determining the Taylor Polynomial Expression

1. For $P_1(x) = T(x) = f(a) + f'(a)(x - a)$, complete the following.
   a. What is $P_1(a)$?
   b. Determine an expression for $y = P_1'(x)$.
   c. What is $P_1'(a)$?
   d. At $x = a$, how do the functions $y = f(x)$ and $y = P_1(x)$ compare? How do the first derivatives of these functions compare at $x = a$?

2. Let P_2 be a general second-degree polynomial:

   $$P_2(x) = b + c(x - a) + d(x - a)^2$$

   Choose values of b, c, and d so that when $x = a$, the function value and the values of the first and second derivatives of P_2 agree with those of f as follows.
   a. What is $P_2(a)$? Set this value equal to $f(a)$.
   b. i. Determine an expression for the first derivative $y = P_2'(x)$.
      ii. What is $P_2'(a)$? Set this value equal to $f'(a)$.
   c. i. Determine an expression for the second derivative of P_2, $y = P_2''(x)$.
      ii. What is $P_2''(a)$? Set this value equal to $f''(a)$.
   d. Solve the expressions in parts (2a) through (2c) for the values of b, c, or d in $y = P_2(x)$.
   e. Replace b, c, and d in the general formula for the second-degree polynomial P_2 listed above with equivalent terms in f, f', and f''. The resulting expression is the **Taylor polynomial of degree 2** for the function f expanded about $x = a$.
   f. Write $y = P_2(x)$ in terms of $y = P_1(x)$.

3. The polynomial approximation can be extended one degree further by adding a third-degree term, $u \cdot (x - a)^3$, to the formula for $y = P_2(x)$. So,

   $$P_3(x) = P_2(x) + u(x - a)^3$$

   a. Show that $P_3(a) = f(a)$.
   b. i. Determine an expression for $y = P_3'(x)$.
      ii. What is $P_3'(a)$?
   c. i. Determine an expression for $y = P_3''(x)$.
      ii. What is $P_3''(a)$?
   d. i. Determine an expression for $y = P_3'''(x)$.
      ii. For the cubic polynomial, it is also necessary that $P_3'''(a) = f'''(a)$. Determine an expression for u in terms of $f'''(a)$.
   e. From parts (3a) through (3d), determine an expression for the **Taylor polynomial of degree 3** for the function f about $x = a$.

If parts (1) through (3) have been completed correctly, students will notice that a pattern is forming. Continuing in this manner, it is possible to determine a Taylor polynomial P_n (an approximating polynomial) of any degree n having the characteristic that the first n derivatives of P_n and f are equal at $x = a$. Following is a general Taylor polynomial of degree n that approximates $y = f(x)$ on an interval containing $x = a$:

$$P_n(x) = f(a) + f'(a)(x - a) + \frac{f''(a)}{2!}(x - a)^2 + \ldots$$
$$+ \frac{f^{(n-1)}(a)}{(n-1)!}(x - a)^{n-1} + \frac{f^{(n)}(a)}{n!}(x - a)^n$$

Written in summation notation, the **Taylor polynomial of degree n** for the function f about $x = a$ is

$$P_n(x) = \sum_{k=0}^{n} \frac{f^{(k)}(a)}{k!}(x - a)^k$$

21.3 Approximating Functions with Taylor Polynomials

1. To see that $y = P_2(x)$ better approximates $y = f(x)$ than $P_1(x) = T(x)$, complete the following.

    a. Let $f(x) = e^x$. Find expressions for P_1 and P_2 expanded around a = 0 using the general expressions determined in Section 21.2 parts (1) and (2). (Determine expressions for $y = f'(x)$ and $f''(x)$, and replace x with 0 in these expressions.)

    b. Graph f, P_1, and P_2 on a viewing rectangle [–2, 2] by [0, 3]. Sketch and label the graphs and axes.

    c. How do the graphs of P_1 and P_2 compare with the graph of f?

    d. Evaluate f, P_1, and P_2 for x equal to 0.2, 0.1, 0.05, and 0.025. Which of P_1 and P_2 provides the best estimate for f?

2. To see how the error changes as the approximation changes on a small interval around a = 0, complete the following.

    a. Graph $y = f(x) - P_1'(x)$ and $y = f(x) - P_2(x)$ on [–.5, .5] by [–.02, .02].

    b. The function $y = P_1$ is linear. What polynomial shape does the shape of the graph of $y = f(x) - P_1(x)$ resemble?

    c. The function $y = P_2$ is quadratic. What polynomial shape does the graph of $y = f(x) - P_2(x)$ resemble?

d.   Suppose that a third-degree polynomial was also used to estimate $y = f(x)$ near $x = 0$. What polynomial would you expect the error curve $y = f(x) - P_3(x)$ to resemble? Why?

3.   Determine expressions for $y = P_3(x)$ and $y = P_4(x)$ for $f(x) = e^x$ expanded around $a = 0$.

4.   a.   Graph the Taylor polynomials P_1, P_2, P_3, and P_4, and $f(x) = e^x$ on a viewing rectangle $[-2, 2]$ by $[0, 3]$. Sketch the graphs using different colors for each graph. Label the graphs and the axes.

      b.   How do the graphs of P_1 through P_4 compare with the graph of f?

      c.   As n grows large, how would you expect the graph of P_n to compare with the graph of f?

5.   a.   As in part (2), sketch the error curves $y = f(x) - P_k(x)$ for $k = 1, 2, 3$, and 4. A first choice for an appropriate viewing rectangle is $[-.5, .5]$ by $[-.02, .02]$.

      b.   Graph these curves on smaller windows until it is possible to distinguish each from the x and y axes. Sketch the curves and label the graphs for the most appropriate window.

      c.   Determine the shape of the error graphs as compared with polynomial functions.

      d.   Which polynomial would you expect the shape of the error curve $y = f(x) - P_5(x)$ to resemble?

6.   a.   Sketch $y = f(x) - P_4(x)$ and the fifth-degree term of the Taylor polynomial $P_5(x)$, $y = \dfrac{f^{(5)}(0)}{5!}(x - 0)^5$, on an appropriate window that clearly shows each of these functions for x near 0. (Choose a very small interval for y.)

      b.   From your sketch in part (6a), is

$$|f(x) - P_4(x)| = \left| \frac{f^{(5)}(0)}{5!}(x - 0)^5 \right| ?$$

      c.   Explain the relationship between these functions.

7.   Repeat the investigation in parts (1) through (6) for $f(x) = \sin x$ expanded around $a = 0$. What do you notice about some of the polynomials $y = P_k(x)$ approximating $f(x) = \sin x$?

8.   The function $y = f(x)$ can be thought of as the sum of a Taylor polynomial P_n and the error R_n introduced by using P_n to approximate f, $f(x) = P_n(x) + R_n(x)$.

      Considering the observations made in parts (5) and (6), for the functions $f(x) = e^x$ and $f(x) = \sin x$.

      a.   What degree would you expect $y = R_n(x)$ to be?

      b.   How might you estimate the size of the error (remainder) term R_n?

21.4 Related Problems

1. a. Determine the first 4 Taylor Polynomials for $f(x) = \ln(x + 1)$ expanded around a = 0. Show your work.

   b. Write $\ln(x+1)$ as an infinite series by generalizing your work from part (1a).

   c. For what values of x does the infinite Taylor Series for $f(x) = \ln(x + 1)$ converge? Use an infinite series test to prove your work.

   d. Suppose your calculator uses a Taylor Polynomial for $f(x) = \ln(x + 1)$ expanded about a = 0 to estimate the values of ln t for any value of t in its radius of convergence. Assuming that the calculator uses only the number of significant digits visible in its display (assuming there are no hidden digits available that the calculator does not display), what is the degree of the Taylor Polynomial that is used to estimate ln 0.9 to the accuracy shown? Show all of your work. Explain your results.

   e. Is it likely that your calculator uses Taylor Polynomials expanded around $x = 0$ to approximate $f(x) = \ln(x + 1)$? Explain in terms of your results in part (1d).

2. a. Show that $e^x = \displaystyle\sum_{k = 1}^{\infty} \frac{x^k}{k!}$ by generalizing your work from Section 21.3 parts (1) and (3).

   b. Show that $e^x = \displaystyle\sum_{k = 1}^{\infty} \frac{x^k}{k!}$ converges for all x

      i. intuitively, using a graphic or a numeric approach.
      ii. formally, using an analytic argument.

3. a. Determine the (infinite) Taylor series expansion for $f(x) = \cos x$ expanded about $x = 0$.

   b. For which values of x does this series converge? Explain.

22

Conic Sections

22.1 Introduction

The purpose of this project is to develop a broader familiarity with conic sections. Conic sections arise from the context of slicing a double cone with a plane. The cone is formed by revolving one line about another line. Conic sections are planar curves whose graphs are equations of the form

$$Ax^2 + Bxy + Cy^2 + Dx + Ey + F = 0.$$

Conic sections have interesting reflective properties which make them useful in a variety of applications. These practical uses of conic sections are also explored.

22.2 Circles

In the general form of the quadratic equation provided in Section 22.1, suppose $A = C \neq 0$ and $B = 0$.

1. Show that the equation above can be written in the standard form for the equation of a circle $(x - h)^2 + (y - k)^2 = r^2$.

2. Are all equations $Ax^2 + Bxy + Cy^2 + Dx + Ey + F = 0$, where $A = C \neq 0$ and $B = 0$, circles? Explain. Determine any restrictions on D, E, and F for which this equation is not a circle.

3. Is the condition $B = 0$ necessary if a general quadratic equation is to define a circle? Explain.

4. Determine the slope of a tangent to the circle $x^2 + y^2 = r^2$ through a point (a, b) on the circle. What is the relationship between the tangent and the diameter of a circle? Explain.

22.3 Parabolas

Definition: A **parabola** is the set of all points P in the plane lying equidistant from a fixed line *l* and a fixed point F.

1. Using the geometric definition of a parabola, how might a parabola be sketched, using a compass and straightedge, if only the point F (the focus) and the line *l* (the directrix) are given? Use your method to sketch a parabola.

2. a. From the definition of a parabola, derive the equation for a parabola in standard position (vertex at the origin). Use the sketch to the right. Show your work.

   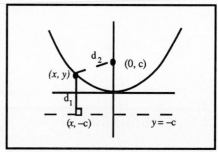

   b. Determine the focus and the equation of the directrix for the parabola $y = x^2$. Explain your work in terms of the derivation completed in part (2a).

   c. Suppose a parabola in standard position has a vertex at the origin and focus on the positive *x*-axis. What is the standard form of the equation of such a parabola? Sketch such a parabola. Label the focus, directrix, vertex, and axis of symmetry.

3. For each of the parabolas given,
   a. sketch the graph of the parabola,
   b. complete the following table, and
   c. explain your work in terms of the derivation completed in part (2), the definition of the parabola, and your knowledge of families of functions.

parabola	focus	directrix	vertex	axis of symmetry
$y = x^2$				
$y = x^2 + 2$				
$y = (x - 3)^2$				
$y = (x - 3)^2 + 2$				
$y = x^2 + 6x - 3$				
$y = -2x^2$				
$y = -2x^2 + 2$				
$y = -2(x - 3)^2$				
$y = -2(x - 3)^2 + 2$				
$y = -2x^2 + 6x - 3$				

(continued)

parabola	focus	directrix	vertex	axis of symmetry
$x = y^2$				
$x = y^2 + 2$				
$x = (y - 3)^2$				
$x = -(y - 3)^2 + 2$				
$x = -y^2 - 6y + 3$				

4. a. What is the formula for the parabola with directrix parallel to the *x*-axis whose vertex is the point (h, k)?

   b. Determine the vertex, focus, directrix, and axis of symmetry of this parabola. Explain your work in terms of families of functions.

5. Use the sketch provided to determine the reflective properties of parabolas. Assume that PR is parallel to the axis of symmetry of the parabola, PQ is tangent to the parabola at point P, F is the focus, and $y = -c$ is the directrix.

   a. Consider the parabola in standard position whose equation was determined in part (2a). Write the parabola as a function $y = f(x)$.

   b. Determine $f(a)$ and rewrite both coordinates of P (a, b) in terms of a.

   c. Determine the slope of the tangent line through point P.

   d. Prove that PQ ⊥ FR.

   e. Use part (5d) and similar or congruent triangles to prove that $\alpha = \beta$.

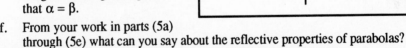

   f. From your work in parts (5a) through (5e) what can you say about the reflective properties of parabolas?

6. Use the reflective property of parabolas to complete the following.

   a. Suppose the graph of a parabola $y = Ax^2$ is revolved about the *y*-axis. By truncating such a figure, you get a satellite dish shape. Determine the placement of a radio/video receiver to receive programs transmitted from satellites in space. Assume that rays entering the Earth's atmosphere are parallel to each other and that the dish is positioned so that rays enter the dish parallel to the axis of symmetry.

   b. Design a round headlamp to fit an automobile. The diameter of the opening is 6 inches, the depth of the receptacle is 2.5 inches. Determine an equation for a parabola in standard position that is to be revolved about the *y*-axis to obtain such a headlamp. Determine the location of the bulb.

22.4 Ellipses

Definition: An **ellipse** is the set of all points in the plane, the sum of whose distances from two fixed points (the foci) is a constant.

1.   Using the geometric definition of an ellipse, how can you sketch an ellipse using only a compass and a straightedge, if the two foci are given? Sketch an ellipse using the method you describe.

2.   For each of the following, sketch each ellipse, determine the major axis, the minor axis, the foci, the center of the ellipse, and the intersection points of the ellipse with the major and minor axes.

   a. $\dfrac{x^2}{4} + \dfrac{y^2}{9} = 1$

   b. $\dfrac{x^2}{9} + \dfrac{y^2}{4} = 1$

   c. $\dfrac{y^2}{4} + \dfrac{x^2}{9} = 1$

   d. $\dfrac{y^2}{4} + \dfrac{x^2}{9} = 1$

3.   An ellipse in standard position is sketched below. The standard form for the equation of the ellipse with the center at $(0, 0)$ is

$$\frac{x^2}{a^2} + \frac{y^2}{b^2} = 1.$$

   a.   Describe the shape and position of an ellipse in which $a < b$.

   b.   Describe the shape and position of an ellipse in which $a > b$.

   c.   Using the family of function ideas, determine the standard form of the equation of the ellipse with center at (h, k).

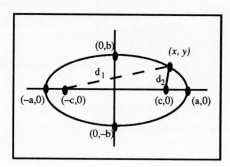

4.   For a general second-degree equation in two variables,

$$Ax^2 + Bxy + Cy^2 + Dx + Ey + F = 0$$

   when $B = 0$, determine the values of A and C for which this expression is the equation of an ellipse.

5.   The eccentricity e of the ellipse is defined as the ratio of $\dfrac{c}{a}$ when $2c$ is the focal length and $2a$ is the length of the semi-major axis.

a. For the standard form for the equation of the ellipse with center at $(0, 0)$,
    $\frac{x^2}{a^2} + \frac{y^2}{b^2} = 1$, how are a, b, and c related? Use the figure provided in part
    (6).

b. Show that $0 \le e < 1$ for every ellipse.

c. Describe the shape of an ellipse for which the eccentricity approaches 0.

d. Describe the shape of an ellipse for which the eccentricity approaches 1.

e. Is a circle an ellipse? Explain.

6. Use the sketch provided to determine the reflective properties of ellipses.
    Assume that PQ_1 is tangent to the ellipse at point P, F_1 and F_2 are the foci,
    and both F_1Q_1 and F_2Q_2 are perpendicular to the tangent to the ellipse.

a. Using the equation of the
    ellipse, write $y = s$ in terms of
    $x = r$. Solve the equation of the
    ellipse for y and rewrite both
    coordinates of P (r, s) in terms
    of r.

b. Determine the equation of the
    tangent to the ellipse

    $$\frac{x^2}{a^2} + \frac{y^2}{b^2} = 1.$$

    through the point P (r, s).

c. Use the following result from trigonometry to show that $\alpha = \beta$.
    For two intersecting lines, l_1 and l_2 with slopes m_1 and m_2 and included

    angle θ, $\tan \theta = \left| \frac{m_2 - m_1}{1 + m_2 m_1} \right|$.

d. In light of the result of part (6c), what can you say about the reflective
    properties of an ellipse?

e. Describe at least two uses that could be made of the reflective properties of
    ellipses.

22.5 Hyperbolas

Definition: A **hyperbola** is the set of all points in a plane, the difference of
whose distances from two fixed points is a constant.

1. Using the geometric definition of a hyperbola, suggest a method that you
    might use to sketch a hyperbola using only a straightedge and a compass.

2. A hyperbola in standard position is sketched below. The standard form for the equation of the hyperbola with center at $(0, 0)$ is

$$\frac{x^2}{a^2} - \frac{y^2}{b^2} = 1, \text{ where } b = \sqrt{c^2 - a^2}$$

a. Determine the equations of the asymptotes for the standard hyperbola using calculus. Show your work.

b. Sketch the graph of each of the following hyperbolas:

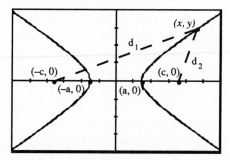

i. $\dfrac{x^2}{4} - \dfrac{y^2}{9} = 1$

ii. $\dfrac{x^2}{9} - \dfrac{y^2}{4} = 1$

iii. $\dfrac{y^2}{4} - \dfrac{x^2}{9} = 1$ iv. $\dfrac{y^2}{9} - \dfrac{x^2}{4} = 1$

c. For each hyperbola, determine the asymptotes, the foci, the center of the hyperbola, and the intersection points of the hyperbola with the x or y axes (whichever is appropriate). Compare the results.

d. Consider the standard form of the equation of the hyperbola given above. For the cases below, describe the shape and position of a hyperbola. Explain.

i. $a < b$ ii. $b < a$

e. Suppose the equation of the hyperbola is $\dfrac{y^2}{a^2} - \dfrac{x^2}{b^2} = 1$ for $b = \sqrt{c^2 - a^2}$. Repeat parts (2a) and (2d) for this hyperbola family.

3. a. Use the family of functions ideas to determine an equation for the standard form of the equation of the hyperbola with center at (h, k).

b. If the equations for the asymptotes for the hyperbola in part (2) are $y = \pm \dfrac{b}{a} x$, what might you expect the equations for the asymptotes to the hyperbola in part (3a) to be? Explain.

4. For a general second-degree equation in two variables,

$$Ax^2 + Bxy + Cy^2 + Dx + Ey + F = 0$$

when $B = 0$, determine values of A and C for which this expression is the equation of a hyperbola.

5. a. Describe the reflective properties of the hyperbola.

b. Describe at least one use that could be made of the reflective properties of a hyperbola.

6. The graph of $y = \dfrac{1}{x}$ is an hyperbola. Determine the asymptotes and foci and show that the geometric definition of the hyperbola is satisfied.

22.6 Rotated Conics

1. Complete the following for each of the quadratic equations $x^2 + Bxy + y^2 = 9$ for values of B = 0, 1, 1.5, 2, 2.5, 3, 3.5, and 4.
   a. Sketch the graph.
   b. Complete the table below.

B	type of curve	B^2-4AC	x-intercept(s)	y-intercept(s)
0	circle	−4	(−3, 0), (0, 3)	(0, −3), (0, 3)
1				

   c. Determine if each is a conic section or a degenerate form that cannot be a slice of a cone by a plane.
   d. Characterize each conic section by the value of its discriminant.
   e. Explain whether or not the xy-term affects the x- and y-intercepts.
2. Repeat part (1) for the quadratic equations $x^2 + Bxy + y^2 - 10x - 8y + 16 = 0$ for the values of B = 0, 1, 2, and 4.
3. Repeat part (1) for the quadratic equation $x^2 + 3xy + Cy^2 + x = 2$ where C = −2.5, −2, −1, 0, 1, 2, and 3.
4. Summarize the predictive properties of the discriminant in determining types of conic sections from the general quadratic equation as follows.
   a. Determine what values for $B^2 - 4AC$ give each of the following.
      i. circle   ii. parabola   iii. ellipse   iv. hyperbola
   b. In terms of the values of A, B, and C, suggest why the conic sections have the values of the discriminants as suggested in part (4a).
   c. For each of the types of graphs found in parts (1) through (3):
      i. How can the graph be obtained by slicing a double cone by a plane?
      ii. Suggest whether or not each graph is a conic section.
      iii. What other kinds of graphs can be obtained by slicing a cone with a plane?
      iv. What values of B in the equations from parts (1) and (2), and what values of C in part (3), will give these conic slices? Explain.

23

Parametric Equations

23.1 Introduction

In Calculus I we were interested in an object's height over time. Suppose an object is launched in a direction that is not perfectly vertical. It might now be interesting to know the object's distance, not only from the ground, but also from the launch site. How does an object travel when launched horizontally? What is its distance from the ground? What is its distance from the launch site? What path does it take? What is its horizontal velocity? What is the vertical velocity? How far does the object travel? Where will it land?

In this activity, students observe the rate at which objects fall and the rate at which objects move horizontally when launched. Students experiment with equations that can be sketched on the x and y-axes, but for which x and y are functions of a third variable. Such equations are called parametric equations. The graphs of parametric equations are called parametric curves.

23.2 Horizontal Versus Vertical Position

Experiment with falling objects as follows. Two objects of the same size, weight and shape, and a third object that is either lighter than or heavier than the others are needed to complete the activity. Choose objects for which wind resistance will be negligible when the objects are thrown only a short distance. (Do not use balloons, feathers, sheets of paper, etc.) Bean bags or plastic bags containing a moderate amount of macaroni work well.

1. Use two objects of same size, weight, and shape. Use a meter stick to guide the height at which objects are dropped or thrown. Stand next to the meter stick with a partner. Drop one of the objects from a height of one meter. At the same time, your partner must gently throw the other object horizontally from

the same height. Both objects **must** start their descent from the same height. The objects cannot be thrown with any initial upward or downward velocity, only horizontal velocity. Repeat the experiment several times. Answer the following.

    a.   Which object lands first?

    b.   Does the horizontal motion of an object affect its vertical motion?

2.   Using the two like objects, simultaneously launch these from the same height, each with a different horizontal velocity. Repeat the experiment several times. Answer the following.

    a.   Which object lands first?

    b.   Does the initial horizontal velocity affect the objects' vertical positions?

    c.   Does the initial horizontal velocity affect the objects' horizontal positions?

3.   Using two different weight objects, repeat the experiments in parts (1) and (2). Answer the questions pertaining to these experiments.

4.   Summarize your results. Explain in terms of the observations made above.

    a.   On what does the horizontal position x of an object depend?

    b.   On what does the vertical position y of an object depend?

5.   The parametric equations that describe the situation above are $x(t) = v_0 t$ and $y(t) = s_0 - 4.9t^2$ where t is time, and x and y are in meters.

    a.   From the results in part (4), state what each of the quantities x, y, v_0, s_0, and -4.9 represent.

    b.   Plot the curve defined by the points $(x(t), y(t)) = (v_0 t, s_0 - 4.9t^2)$.[1] Describe the curve in terms of a familiar function $y = f(x)$.

    c.   Write the parametric form of the curve as a function $y = f(x)$ which gives the position y with respect to x directly. Explain the process you used.

    d.   What do x and y represent in the equation in part (5c)?

    e.   What information is provided by the function $y = f(x)$? How does this information relate to the situation explored?

    f.   What additional information is provided by the parametric equations which model the height of an object launched horizontally? Explain how this information is different from that provided by the function $y = f(x)$.

6.   a.   In the parametric form of the curve, x and y each describe positions. For the parametric equations in part (5), determine expressions for $x'(t)$ and $y'(t)$. What do these quantities represent in terms of the situation described and the graph sketched?

    b.   Determine $y = f'(x)$ for the function determined in part (5c). What does f' represent in terms of the situation and the graph?

---

[1] Program PARAMETRIC is provided in Appendix A for CASIO 7000 users. PARAMETRIC plots parametrically defined curves. CASIO 7700 and TI-81 users can plot parametric curves using the parametric graphing mode of the calculator.

    c. How is the equation for f' related to the expressions for x' and y'? Explain.

7. a. When an object is launched horizontally, where will it strike the ground? Explain.

    b. How far will the object travel from the moment it is launched to the moment it lands? Explain.

    c. Explain the difference between questions (7a) and (7b).

23.3 Parametric Curves

In Section 23.2, the usefulness of parametrically defined curves was investigated. In this section it will be shown that the same curve can have several different parametric definitions. Further, curves which cannot be defined as functions $y = f(x)$ might be defined parametrically in terms of functions $x = x(t)$ and $y = y(t)$. The equations $x = x(t)$ and $y = y(t)$ are called parametric equations. The variables x and y are given as functions of a third variable t, such as $(x, y) = (x(t), y(t))$.

1. a. Graph the curves defined by the following ordered pairs on a viewing rectangle $[-2\pi, 2\pi]$ by $[-3, 3]$ for t in $[-\pi, \pi]$.

      i. $(t, \sin t)$ ii. $(t, 2\sin t)$ iii. $(t, \sin 2t)$
    iv. $(2t, \sin t)$ v. $(2t, \sin 2t)$ vi. $(2t, 2\sin t)$

    b. For each parametric curve in part (1a), write x as a function of t. Write y as a function of t.

    c. Which pair(s) of parametrically defined curves in part (1a) seem to sketch exactly the same curve?

    d. Trace along the curves found in part (1c). Complete the following table for each of these curves. Describe any similarities and/or differences.

t	$-\pi$	$-\dfrac{\pi}{2}$	$-\dfrac{\pi}{4}$	0	$\dfrac{\pi}{4}$	$\dfrac{\pi}{2}$	π
$x(t)$							
$y(t)$							

    e. For the pair(s) of parametric curves described in part (1d), suppose two objects are moving along the curve. If the movement of both objects started at time $t = 0$, describe the change in distance between these objects as time goes on.

    f. Experiment with the interval for t over which the parametric curves in part (1a) are sketched. Determine the smallest interval for t that will guarantee a range of $[-2\pi, 2\pi]$ for y for each curve.

g. From the graphs sketched in part (1a) write the equations for the parametrically defined curves in terms of the function $y = f(x)$. Explain your results.

h. Show analytically that your results in part (1g) are correct.

2. a. Graph the curves defined by the following ordered pairs on the viewing rectangle $[-5, 5]$ by $[-3, 3]$ for t in $[-2\pi, 2\pi]$. Clear the graphics screen after each graph. Describe any similarities and/or differences in the graphs.

    i. $(|t|, \sin |t|)$ ii. $(\sqrt{t}, \sin\sqrt{t})$ iii. $(t^2, \sin t^2)$

    iv. $(\cos t, \sin (\cos t))$ v. $(\tan t, \sin (\tan t))$ vi. $(\frac{1}{t}, \sin \frac{1}{t})$

  b. Describe the ranges for x and y. What effect does the choice of x in the parameterization of $y = \sin x$ have on the graph?

  c. Choose a parameterization $(x(t), y(t))$ of $y = \sin x$ so that only the principal branch of the sine function is sketched.

  d. From part (2c), suggest parametric equations $x(t)$ and $y(t)$ to sketch $y = \text{Sin}^{-1}(x)$ using the sine function.

3. a. Sketch the curves defined by the following ordered pairs on the viewing rectangle $[-5, 5]$ by $[-3, 3]$ for t in $[0, 2\pi]$. Describe each graph in terms of familiar curves. Describe any similarities and/or differences in the graphs.

    i. $(\cos t, \sin t)$ ii. $(2\cos t, 2\sin t)$ iii. $(\cos t^2, 2\sin t^2)$

    iv. $(\cos t, \cos 2t)$ v. $(\sec t, \tan t)$

  b. Do each of the graphs sketched in part (3a) define a function $y = f(x)$? Explain.

  c. For each parametric curve sketched in part (3a), write x as a function of t. Write y as a function of t.

  d. What are the equations of the functions and/or relations sketched in part (3a), where y is written in terms of x? Determine these from the graphs.

  e. For each parametric curve in part (3a), choose the smallest interval for t to sketch the curve only once.

  f. Sketch the parametric curves defined below. Clear the graphics screen after each graph. Describe any similarities and/or differences in the graphs.

    i. $(\cos 2t, \sin 2t)$ ii. $(\cos t, \sin (-t))$

  g. How do the graphs of the curves sketched in part (3f) compare with those sketched in part (3a)?

  h. Write the parametrically defined curves in parts (3a) and (3e) as equations in x and y. Show your work. (Use trigonometric identities.)

4. For each of the standard conic sections listed below, suggest a parameterization to graph each using trigonometric functions.

  a. Circle, $x^2 + y^2 = r^2$ b. Parabolas, $y = ax^2$ and $x = ay^2$

  c. Ellipse, $\dfrac{x^2}{a^2} + \dfrac{y^2}{b^2} = 1$

d. Hyperbolas, $\dfrac{x^2}{a^2} - \dfrac{y^2}{b^2} = 1$ and $\dfrac{y^2}{a^2} - \dfrac{x^2}{b^2} = 1$

e. Comment on any restrictions to the domain and/or range of each of the conic sections above when parameterized using trigonometric functions for x and y.

5. How can the parameterizations of the conics in part (4) be adjusted to graph the following general conic sections? Explain.

   a. Circle, $(x - h)^2 + (y - k)^2 = r^2$

   b. Parabolas, $(y - k) = a(x - h)^2$ and $(x - h) = a(y - k)^2$

   c. Ellipse, $\dfrac{(x - h)^2}{a^2} + \dfrac{(y - k)^2}{b^2} = 1$

   d. Hyperbolas, $\dfrac{(x - h)^2}{a^2} - \dfrac{(y - k)^2}{b^2} = 1$ and $\dfrac{(y - k)^2}{a^2} - \dfrac{(x - h)^2}{b^2} = 1$

6. Parameterizations of functions do not need to be written in terms of trigonometric functions. By now you have discovered that infinitely many possibilities exist. Consider the following parameterizations. How do these relate to the line l through the point $(1, 2)$ with slope of -1 for t in $(-\infty, \infty)$?

   i. $x(t) = t$ $\qquad\qquad$ ii. $x(t) = 2t$
       $y(t) = -t + 3$ $\qquad\qquad\quad$ $y(t) = -2t + 3$
   iii. $x(t) = t^2$ $\qquad\qquad$ iv. $x(t) = \cos t$
       $y(t) = -t^2 + 3$ $\qquad\qquad$ $y(t) = -\cos t + 3$

   a. Explain how each pair of equations was determined.

   b. Graph each parametrically defined curve. Does each pair of equations define the same curve? Explain. Note any similarities and/or differences. (Clear the graphics screen before sketching each curve.)

   c. Complete the table below for the parametric equations in part (6) for t in $(-\infty, \infty)$.

$(x(t),\ y(t))$	x Domain	x Range	y Domain	y Range
$(t, -t + 3)$				
$(2t, -2t + 3)$				
$(t^2, -t^2 + 3)$				
$(\cos t, -\cos t + 3)$				

   d. Is it possible for the parametric equations given in part (6) to provide the same graph? Explain in terms of the domains and ranges for x and y in each instance.

23.4 Calculus of Parametric Curves

The connections between parametric curves and equations in x and y were explored in previous sections of this investigation. In Section 23.4, the calculus of functions

$y = f(x)$ is extended to parametric curves. These ideas were alluded to in Section 23.2 when the derivatives of $y = f(x)$, $x(t)$ and $y(t)$ were considered. They are discussed more completely here.

Recall the development of the derivative from Calculus I.[2] The relationship between rate of change from a police officer's perspective versus a commuter's perspective was discussed then modeled graphically. Graphically, the commuter's velocity could be determined by finding the slope of a line segment containing the points $(a, f(a))$ and $(b, f(b))$ where a and b are beginning and ending times and f is the commuter's position. The police officer's view was at the time when the commuter was exceeding the speed limit, and only for a short time, say from time c to time c + h, where h is a very small number. This view is modeled below.

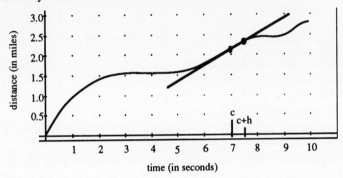

An expression for velocity from the police officer's point of view at time $t = c$ was determined to be

$$v(c) = f'(c) = \lim_{h \to 0} \frac{f(c + h) - f(c)}{h}$$

1. Suppose the curve above is the curve defined by parametric equations $x(t)$ and $y(t)$.

    a. Label two points $(x(t_0), y(t_0))$ and $(x(t_0 + h), y(t_0 + h))$. Determine the slope of the line containing these points.

    b. Write the quotient from part (1a) as a limit showing the change in y divided by the change in x as the time interval from t_0 to $t_0 + h$ approaches zero.

    c. Note that in the discussion above, the limit for the commuter's velocity is written in terms of a quotient giving the change in the dependent variable with respect to the change in the independent variable. In the quotient in part (1b) both x and y are dependent variables. To use the derivative definition to define the slope of the parametric curve in terms of derivatives of x and y, it will be necessary to write the quotient in part (1b) in terms of the derivative of x and the derivative of y. How might this be done?

---

[2] See Chapter 6: *Understanding the Derivative* for a review.

Write this quotient and show that it is equivalent to the expression found in part (1b).

d. What is an expression for the slope of a tangent to a parametric curve $(x(t), y(t))$ at $t = t_0$ in terms of the derivatives of x and y?

e. For the expression found in part (1d), interpret the following cases geometrically.

   i. $x'(t) = 0$ when $y'(t) \neq 0$ ii. $y'(t) = 0$ when $x'(t) \neq 0$

  iii. $x'(t) = 0$ and $y'(t) = 0$

2. The curve $(x(t), y(t))$ is given in parametric form.

$$x(t) = 3 \cos t$$

$$y(t) = 4 \cos 2t$$

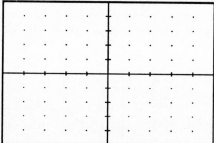

a. Sketch the curve on the axes at right. Label the x and y scales on the axes.

b. Use the parametric form of the equation to determine all the values of t for which the curve has a horizontal tangent. Explain your work.

c. Determine the slope of any tangents for which the slope is undefined when determined from the parametric form of the curve. Describe the behavior of $(x(t), y(t))$ at these points and explain why the parametric derivatives are not helpful in determining the slope of the tangent at these points.

d. What familiar function does this graph resemble?

e. Determine an expression in terms of $y = f(x)$ for the function given in parametric form above.

f. Are the graphs of the parametric curve $(x(t), y(t))$ and the function $y = f(x)$ exactly the same? Describe any differences, and explain why they occur.

g. Determine the equation of the tangent to the graph of $y = f(x)$ at $x = 1$. Explain your work.

h. In terms of the parametric equations given above, for what value of t is the tangent in part (2g) found?

3. Consider the parameterizations of the function $y = x^3 + 1$.

   i. $x(t) = t$ ii. $x(t) = |t|$

      $y(t) = t^3 + 1$ $y(t) = |t|^3 + 1$

  iii. $x(t) = \cos t$ iv. $x(t) = \sqrt{t}$

       $y(t) = (\cos t)^3 + 1$ $y(t) = \sqrt{t^3} + 1$

a. Sketch each of the parametric curves on the viewing rectangle [–2, 2] by [–8, 8] for t in [–1, 0].[3] Record the results on paper using the same sized grid (as shown). Clear the screen after each sketch.

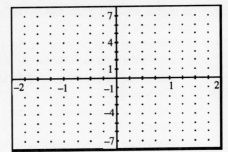

b. Repeat part (3a) for t in [1, 2].

c. Describe the direction each graph is drawn, comparative positions of graphs, and comparative lengths of the arcs drawn.

d. Does the choice of parameterization of the function $y = x^3 + 1$ affect the length of the curve sketched over a t-interval [a, b]? Explain.

e. For any non-constant function $y = f(x)$, would you expect that the parameterizations $(x(t), f(x(t)))$ to result in different curve lengths for t in [a, b]? Explain.

4. Determining the length of an arc is a basic application of the distance formula and the definition of the definite integral.

a. Suggest an approach to determine the length of an arc using the distance formula and the definition of the definite integral. Do not derive the formula. Discuss the basis for the formula and why it makes sense.

b. The arc length L of a parametrically defined curve $(x(t), y(t))$ over the interval [a, b] is given by the formula

$$L = \int_a^b \sqrt{(x\,'(t))^2 + (y\,'(t))^2}\, dt$$

Determine the arc lengths of the curves sketched in part (3a). Set up the integrals and use numerical integration techniques to estimate the length of the arc. Complete the table below. Explain your results.

$(x(t),\ y(t))$	arc length over [–1,0]	arc length over [1, 2]
$(t,\ t^3 + 1)$		
$(|t|,\ |t|^3 + 1)$		
$(\cos t,\ (\cos t)^3 + 1)$		
$(\sqrt{t},\ \sqrt{t}^3 + 1)$		

5. Reconsider the problem in Section 23.2. When an object is launched with an initial horizontal velocity of v_0, the parametric equations that describe the horizontal distance of the object from the launch site and the vertical height of

---

3 Use a t-step of .02 to obtain a reasonably accurate graph within a reasonable amount of time.

the object are $x(t) = v_0t$ and $y(t) = s_0 - 4.9t^2$ where t is time, s_0 is the initial height and x and y are in meters.

a. Determine an expression for the distance the object travels from the moment it is launched until it strikes the ground.

b. Determine an expression for the velocity at which the object is traveling horizontally.

c. Determine an expression for the velocity at which the object is traveling vertically.

d. What does the slope of the tangent to the parametric curve represent in this situation?

e. Can the information from parts (5b) and (5c) be obtained from the information in part (5d) if the equation for the curve is given as a function $y = f(x)$ rather than parametrically as $(x(t), y(t))$? Explain.

6. Suppose an object is launched with an upward initial velocity v_0 and the angle from horizontal at which the object is launched is α. This situation is described by the parametric equations

$$x(t) = (v_0 \cos \alpha)t + x_0 \text{ and } y(t) = -\frac{1}{2}gt^2 + (v_0 \sin \alpha)t + y_0$$

where (x_0, y_0) is the initial point of launch, g is the acceleration due to gravity, and t is time.

a. Compare this parameterization of the curve to that in part (5). How are these parametric curves related?

b. Suppose an airplane is to drop blankets to refugees in a remote mountain valley along the Iraq/Turkey border. Because of the terrain, the airplane must fly horizontally at approximately 0.8 km above the valley. Suppose that the plane is flying with a velocity of 240 km/hr.

   i. How long will it take the blankets to reach the valley floor?

   ii. What horizontal distance will the blankets travel from the moment they are dropped from the plane?

   iii. How far do the blankets travel in the air?

   iv. In order to drop the blankets so that they land 1 km beyond the refugees (but not on the refugees!), at what angle of site between the horizontal and the direct line to the target should the blankets be released so that they land on target?

c. The fireworks at a display are to be set off in the middle of a lake.

   i. At what angle of inclination should the fireworks be set off so that they travel the furthest horizontal distance? Explain.

   ii. At what angle should they be set off so that they travel the furthest vertical distance? Explain.

24

Polar Functions

24.1 Introduction

Polar functions answer the need for a coordinate system that depends on a distance from some central point and the amount of rotation from an initial direction. For example, an air traffic controller is interested in the distance of a plane from the control tower and from what direction the plane is approaching the tower. Controllers in busy airports such as La Guardia in New York and O'Hare in Chicago are also concerned about the altitude at which the 70 or so planes circling the airport are flying! This investigation concerns angle and direction. The third dimension (altitude) is not addressed here.

Polar functions arise in other applications, as well:
- The orbits of planets with respect to a particular location (such as the sun)
- Locations of celestial bodies with respect to a fixed location (perhaps the earth)
- The study of gravitational and magnetic fields
- The direction and force of a spray of water through a sprinkler head on a lawn, particularly on a lawn of unusual shape
- The path of a car on an amusement park ride that travels around a central pole (such as the *Tilt-O-Whirl* or the *Scrambler*)

The location that serves as the center of the coordinate system is called the pole, O. The fixed direction from which all other directions will be measured is called the polar axis. The pole is the origin (0, 0). The polar axis is usually chosen to coincide with the positive x-axis. The polar coordinates P = (r, θ) are the distance r from the pole and the angular measure θ from the polar axis measured in the counterclockwise direction.

In this chapter, investigations consider what kinds of graphs emerge from the family of polar functions defined by the equation $r = A f\left(B\theta - \dfrac{C\pi}{4}\right) + D$ where f is the sine, cosine, constant function, or the function $r = \theta$, or the reciprocal of one of these. The relationship between polar, conic and parametric equations is also investigated.

24.2 Conversions Between Polar and Rectangular Coordinate Systems

1. A point P in the polar coordinate system is indicated by the coordinates (r, θ). This same point in the rectangular coordinate system is indicated by the coordinates (x, y) as shown in the figure at right.

   $P = (r, \theta) = (x, y)$

   a. Label the corresponding parts of the right triangle with the values x, y, r, and θ.

   b. Determine an expression for x and an expression for y in terms of r and θ.

   c. Determine an expression for r and an expression for θ in terms of x and y. The expression for r should allow r to be any real number (positive or negative).

2. The relationships determined in part (1) provide the key to the relationship between polar and parametric equations. It is possible to graph polar functions on the rectangular coordinate plane by graphing the parametrically defined curve $(x(\theta), y(\theta))$.

   a. Write the parametric form $(x(\theta), y(\theta))$ of the polar function $r = 3$. Graph the resulting curve.[1]

   b. Write the parametric form of the polar function $r = f(\theta) = \sin \theta$. Graph the resulting parametric curve.

3. a. Use the relationships between x, y, r, and θ determined in part (1b) and (1c) to determine an equation in polar coordinates in order to graph a line $y = mx + b$ that does not go through the origin.

   b. Write the equations for the lines $y = 2x - 1$ and $y = -.5x + 1$ as polar functions.

   c. Write the parametric form of the polar functions in part (3b).

   d. Sketch each graph from its polar or parametric form.

---

[1] The parametric graphing utility (CASIO 7700 or the TI-81) or program PARAM (CASIO 7000) can be used to graph parametrically defined polar curves.

24.3 Polar Families[2]

1. For the rectangular coordinate system, the graphs of the equations $y = c$ and $x = k$, where c and k are constants, are lines.

   a. In the polar coordinate system, what are the results of graphing the equations $r = c$ and $\theta = k$ where c and k are constants? Explain.

   b. Sketch $r = D$ for $D = 1, 2, 3, -1$, and .5. Describe the graphs in terms of a familiar curve. Explain why the graphs appear as they do.

2. Let $r = \theta$. Consider the family of functions that arise from changes in the parameters A and D in the equation $r = A\theta + D$.

   a. Graph $r = \theta$ on the the viewing rectangle [–14, 14] by [–9, 9] for θ in [0, 2π].

   b. On the same viewing rectangle, graph $r = \theta$ and $r = A\theta$ for A = 2, –1, and 0.5. Describe the graph of $r = A\theta$ as it compares to the graph of $r = \theta$. Clear the screen before each new graph.

   c. On the viewing rectangle [–10, 10] by [–6, 6] graph $r = \theta + D$ for D = 1, –1, 2, and –3. Explain in terms of the change in r why the graph appears as it does.

3. Let $r = \sin\theta$. Consider the family of functions that arise from changes in the parameters A, B, C, and D in the equation $r = A\sin\left(B\theta - \dfrac{C\pi}{4}\right) + D$.

   a. Graph $r = \sin\theta$ on the viewing rectangle [–5, 5] by [–3, 3] for θ in [0, π].

   b. On the same viewing rectangle, graph $r = A\sin\theta$ for A = 2, –1, and $-\dfrac{5}{3}$. Describe each graph as it compares to the graph of $r = \sin\theta$.

   c. Vary the θ-interval, graphing $r = A\theta$ over the intervals $[0, \dfrac{\pi}{2}]$, [0, π], and [0, 2π]. What is the smallest θ-interval over which a complete graph of $r = A\theta$ is drawn? Explain.

   d. Graph $r = \sin B\theta$ for B = –1, 2, 3, 4, 5, $-\dfrac{5}{3}$, and –2 on the viewing rectangle [–2.2, 2.2] by [–1.5, 1.5] for θ in [0, 2π].

      i. Describe each graph as it compares to the graph of $r = \sin\theta$.
      ii. Describe these graphs as they relate to the graph of $r = 1$.
      iii. Explain why $r = \sin B\theta$ appears as it does.
      iv. Graph $r = \sin B\theta$ for B = $\dfrac{1}{2}, \dfrac{2}{3}$, and $-\dfrac{1}{2}$. Experiment with θ–intervals until a complete graph of each curve is drawn. Record the value of θ needed.

---

[2] To get the best view of polar functions, use a range which gives x and y in the same scale. This can be accomplished by using the program SQ.SCREEN on the CASIO or by using the SQUARE option in the ZOOM menu on the TI-81. Use program POLAR included in the CASIO and TI appendices to graph families of polar functions.

v. How many petals will $r = \sin B\theta$ have for B an integer and for B a rational number? Explain. (Use the graph of $y = \sin Bx$ drawn on the rectangular coordinate system to assist in your explanation.)

e. Graph $r = \sin 2\theta$ for the interval values suggested in part (3c). Graph

$r = \sin\left(2\theta - \dfrac{C\pi}{4}\right)$ for C = 1, 2, 3, 4, and –1. Clear the screen before

starting each new sketch. Compare each graph with the graph of $r = \sin 2\theta$. Note any similarities and/or differences. What is the effect of changing C on the graph of this function? Why?

f. Graph $r = \sin\theta$ on a viewing rectangle [–2.4, 2.4] by [–1.2, 2].

i.   For θ in [0, 2π], graph $r = \sin\theta + D$ for D = 1, .5, .2, and –.5.

ii. Graph $r = \sin\theta + D$ on a viewing rectangle [–3, 3] by [–1, 3] for D = 1, –1, 2, and –2.

iii. What effect does the value of D have on the graph of $r = \theta + D$? Explain.

g. Predict the appearance of the graph of $r = 2\sin\left(3\theta - \dfrac{\pi}{4}\right)$. Explain why

you expect the graph to appear as it does. Graph the function.

h. Graph $r = 2\sin\theta - 1$ on the viewing rectangle [–3, 3] by [–1, 3]. Explain why the graph makes sense in terms of the families of polar functions as explored above.

4. Graph $r = f(\theta) = \cos\theta$.

a. How are the graphs of $r = \cos\theta$ and $r = \sin\theta$ related? Explain.

b. In part (3) replace the sine function with the cosine function. Repeat part (3) as much as necessary to determine the individual effects of changes in

A, B, C, and D on the graph of $r = A\cos\left(B\theta - \dfrac{C\pi}{4}\right) + D$.

i.   What effect does the value of A have on the graph of $r = A\cos\theta$? Why?

ii. Explain why $r = \cos B\theta$ appears as it does.

iii. What effect does the value of C have on the graph of

$r = \cos\left(2\theta - \dfrac{C\pi}{4}\right)$? Why?

iv. What effect does the value of D have on the graph of $r = \theta + D$? Explain.

c. Predict the appearance of the graph of $r = 2\cos\left(3\theta - \dfrac{\pi}{4}\right)$. Explain why

you expect the graph to appear as it does. Graph the function. Compare this graph to the one obtained in part (3g). Explain any similarities and/or differences.

d. Graph $r = 2\cos 3\theta - 1$. Explain why the graph makes sense in terms of the families of polar functions as explored above.

5. Graph $r = f(\theta) = \dfrac{1}{\theta}$ on a viewing rectangle [–2, 2] by [–1.2, 1.4] for θ in

[0, 4π].

a. Use limits as θ approaches 0 and as θ approaches ∞ to explain the appearance of the graph of f. Are there any asymptotes for f? Explain.

b. How does the appearance of the graph change for θ in $[-4\pi, 4\pi]$? Why?

24.4 Polar Versions of Conic Sections

1. a. Which of the polar graphs sketched in Section 24.3 are circles? Write the parameterized form of each circle using the parameter A, B, C, or D.

   b. Determine an equation in x and y for each circle whose general form is listed in part (1a).

   c. Determine the centers and radii for each of the circles whose equations were determined in part (1b). Show that these results agree with the polar graphs.

   d. Write the parametric form of each circle in part (1a).

2. Consider the family of functions defined by the polar equation

   $$r = f(\theta) = \frac{AE}{1 + E \cos \theta}.$$

   a. Graph f on a viewing rectangle $[-5, 5]$ by $[-3, 3]$ for A = 1 and E = .5, .75, 1, 1.25, 1.5, and 2.

   b. Describe each conic section sketched in part (2a) in terms of its focus or foci, directrix (if applicable), axis (axes) of symmetry, focal length, and eccentricity.

   c. Let A = 2. Graph f on a viewing rectangle $[-15, 15]$ by $[-10, 10]$ for A = 1 and E = .75, .85, 1, 1.25, 1.5, and 2.

   d. Describe each conic section sketched in part (2c) in terms of its focus (foci), directrix (if applicable), axis (axes) of symmetry, focal length (if applicable), and eccentricity.

   e. From the observations made in parts (2a) through (2d), for what value of E is the graph of $r = f(\theta) = \dfrac{AE}{1 + E \cos \theta}$

   i.   a parabola,   ii.   an ellipse,   iii.   a hyperbola.

   f. Express $r = \dfrac{E}{1 + E \cos \theta}$ as an equation in x and y for E = .5, 1, and 2. Explain why these choices of E are important in the determination of the type of conic section obtained.

3. a. Graph $r = f(\theta) = \dfrac{A}{2 + \cos \left(\theta - \dfrac{C\pi}{4}\right)}$ on a viewing rectangle

   $[-10, 10]$ by $[-6.6, 6.6]$ for θ in $[0, 2\pi]$ where A = 5, and C = 0, 2, 4, and 6.

   b. Describe the shape and position of the resulting graph.

   c.   What is the effect on the graph of f caused by a change in the parameter C?

   d.   Describe the change in the graph of f if A is chosen to be 1, 2, –.5, or –3. Explain the affect of A on the graph of f.

   e.   Replace the cosine function by the sine function and repeat part (3a). How do the resulting graphs compare with those sketched in part (3a)?

   f.   Determine the focus(foci) of the conic section defined by the polar function

$$r = \frac{5}{2 + \cos \theta} \; .$$

24.5 Area of Polar Regions

In Chapter 13: *Determining Distance from Velocity*, we determined that the area under the function $y = f(x)$ on the x-interval [a, b] can be found by partitioning [a, b] into n equal-width subintervals $[x_{j-1}, x_j]$ and by finding the area of smaller regions over each subinterval. For small enough subintervals (n→∞), subregions under the curve for each subinterval are nearly rectangular. Hence, the area of a subregion can be found by multiplying the width of the interval Δx by its height $f(t_j)$, where the change in the height is negligible over small enough subintervals. Thus, the area of a region is the sum of the areas of the subregions as the number of subregions increases to infinity, $A = \lim\limits_{n \to \infty} \sum\limits_{j=1}^{n} f(t_j) \, \Delta x \; = \; \int_{a}^{b} f(x) \, dx.$

The process for finding areas of polar regions is similar to that for finding areas of rectangular regions. For polar functions, the distance from a pole r is dependent on angular direction θ. The regions over which areas are to be found are also given in terms of distance r and angular direction θ.

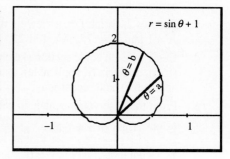

1.   Consider the polar region sketched in the figure at right.

   a.   As with rectangular functions, partition the θ-interval [a, b] into n equal-width subintervals. Describe the shape of the resulting subregion over subinterval $[\theta_{j-1}, \theta_j]$.

   b.   Areas of subregions described in section (1a) can be estimated by using the area of a sector of a circle. Explain why this choice is reasonable.

   c.   Determine the following areas of circular sectors with radius r.

      i.   circle   ii.   semicircle iii.   quarter circle iv.   eighth circle

   d.   What is the area of the circular sector with angular measure $\Delta\theta$?

e. What is the radius of a polar subregion over a very small subinterval? Explain.

f. Write the sum of areas of the subregions of a polar region using Riemann Sum notation.

g. Write the expression found in (1f) as a limit to determine the exact area of the polar region.

h. Use the definition on the definite integral to write the expression in part (1g) as a definite integral.

2. Find the area of the polar region bounded by the curves $\theta = \dfrac{\pi}{6}$, $\theta = \dfrac{\pi}{3}$, and $r = \sin \theta + 1$.

3. Two circles have radius a, and each passes through the center of the other, as shown in the sketch to the right.

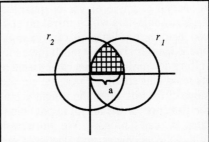

a. Determine polar functions for each of the circles.

$r_1 = f_1(\theta)$ and $r_2 = f_2(\theta)$

b. Determine the parametric form of each function in part (3a).

c. Determine all points of intersection of the graphs of $r_1 = f_1(\theta)$ and $r_2 = f_2(\theta)$. Explain your results.

d. If two particles are traveling around the circles r_1 and r_2 from time $t = \theta = 0$ to time $t = \theta = 2\pi$, do the particles ever collide? Explain.

e. If the particles collide, how far does each particle travel for $t = \theta$ in $[0, \pi)$ to the point where they meet the first time? Show your work. Explain your results.

f. Determine the area of the shaded region in the sketch shown to the right. Set up an integral using polar equations. Explain how you determined the limits of integration. Complete the integration using a numeric method of integration.

g. Is it possible to find the area of the region above using equations given in rectangular coordinates? Explain.

25

Vectors in the Plane

25.1 Introduction

Up until now, our work in calculus has been with quantities that can be represented by a single real number such as time, mass, length, and temperature. Such quantities are called scalars. Vectors are quantities that possess both magnitude and direction. Examples of vector quantities include weight, velocity, and force.

Consider the following situation: A boater is traveling across Lake Michigan from Grand Haven, Michigan to Milwaukee, Wisconsin. She uses a compass to head the boat directly west and sets her speed at 30 mph. A steady 10 mph wind is blowing from the southwest. In what actual direction is the boat heading? Such a problem may be solved with vectors, using one vector to quantify the boat's speed and direction, and a second vector to quantify the wind speed and direction.

A second example of vector quantities comes from San Francisco. Picture the historic cable cars which travel up and down the steep hills of the city. When the conductor brings the cable car to a stop, he releases the car from the cable and pulls on the hand-brake. How much force must the brake exert to keep the cable car from rolling downhill? Vectors can be used to quantify the weight of the cable car due to its mass and gravity. The cable car must travel in the direction of the downward slope of the hill. From the weight vector and the component of the weight vector in the direction of the hill, it is possible to determine what force must be exerted to keep the car stationary. We will revisit these problems as we make sense of the arithmetic, geometry, and algebra of vectors in the plane.

In some applications, it is useful to think of vectors as having a position and a direction. For example, in determining a vector that shows the capital gain of a company from various sources, it makes sense that the vector begin at the origin. In physical applications, it is often best to think of vectors as having a magnitude and a direction, without regard to an initial position. Such vectors are called displacement vectors.

25.2 Arithmetic of Vectors

Definition: The set of **vectors in the plane** is the set of all ordered pairs $\langle a, b \rangle$ of real numbers together with the following rules for addition and multiplication by a real number c:

 i. $\langle a_1, b_1 \rangle + \langle a_2, b_2 \rangle = \langle a_1 + a_2, b_1 + b_2 \rangle$

 ii. $c\langle a, b \rangle = \langle ca, cb \rangle$

The numbers a and b are the **components** of the vector $\langle a, b \rangle$.

1.   Vectors in the direction of the positive x-axis are scalar multiples of the unit coordinate vector $i = \langle 1, 0 \rangle$. Vectors in the direction of the positive y-axis are scalar multiples of the unit coordinate vector $j = \langle 0, 1 \rangle$.

 a.   Using scalar multiplication and vector addition, write the following vectors in terms of the vectors i and j.

 i. $\langle 1, 1 \rangle$ ii. $\langle 2, 3 \rangle$ iii. $\langle 0, b \rangle$ iv. $\langle a, b \rangle$

 b.   Add vectors u and v by using vector ordered pair notation, then by using unit coordinate vector notation.

 i. $u = \langle 1, 1 \rangle$ $v = \langle 2, 1 \rangle$ ii. $u = \langle 0, 3 \rangle$ $v = \langle 2, -1 \rangle$

2.   The magnitude of a vector is its length. Magnitude can be determined using the Pythagorean Theorem or distance formula. The magnitude of a vector v is called the norm of v and is denoted as $|v|$. Vector $v = \langle a, b \rangle$ is sketched below. The magnitude of v is indicated by its length. The direction of v is indicated by the direction that the arrow is pointing.

 a.   What are a and b for the vector sketched?

 b.   Write v as the sum of scalar multiples of the unit coordinate vectors $i = \langle 1, 0 \rangle$ and $j = \langle 0, 1 \rangle$.

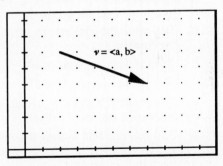

 c.   Sketch the i and j directional components of v on the axes provided.

 d.   Use the Pythagorean Theorem to determine the magnitude $|v|$. Show your work.

 e.   In general, for any vector $v = \langle a, b \rangle$, write an expression for the norm $|v|$.

3.   In applications, it is also useful to name vectors in terms of the angles that they make with the positive x-axis.

 a.   Position the vector $u = \langle 3, 1 \rangle$ so that it originates at the origin and terminates at the point (3, 1).

 b.   Determine the angle α between u and the positive x-axis.

 c.   Determine the length of the vector $\langle \cos \alpha, \sin \alpha \rangle$.

 d.   Write u as a scalar multiple of the vector $\langle \cos \alpha, \sin \alpha \rangle$.

25.3 Geometry of Vectors

The following activities are designed to develop your understanding of the geometric meaning of addition and subtraction of vectors.

1. a. Sketch the following pairs of vectors on graph paper.

>   i. $u = <2, 4>, v = <3, 1>$ ii. $u = <1, 2>, v = <-3, 1>$

   b. Using the definition of vector addition provided in Section 25.2, determine the following for each pair of vectors and sketch the resulting vector. [1]

>   i. $u + v$ ii. $u - v$ iii. $2u + v$ iv. $2u - v$

   c. Suggest a geometric relationship between the sum and the vectors u and v.

   d. Suggest a geometric relationship between the difference of the vectors and the vectors u and v.

2. Consider the boater's problem described on page 144. The figure at right shows vectors that model this situation. Vector b has magnitude 30 in the westerly direction. Vector w has magnitude 10 and is headed in the northeast direction (since the wind is blowing from the southwest).

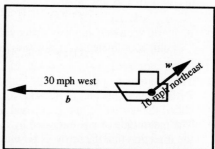

   a. Determine the components of vector b. Note that b is in the direction opposite of the positive x-axis.

   b. Determine the components of vector w. Note that w is in the direction of the line $y = x$.

   c. What is the magnitude of w?

   d. What is the magnitude of b?

   e. As a result of both the wind and the boat speed, in which direction would you expect the boat to travel?

   f. Add vectors b and w using the definition of vector addition. Sketch the vector $v = b + w$ which represents the actual direction and speed at which the boat is traveling.[2]

   g. What does vector v represent geometrically in terms of vectors b and w?

   h. In what direction is the boat traveling? At what speed?

3. The boater in part (2) decided to turn back (lack of fuel, too many sea gulls, not enough sunscreen...). The boat's compass heading indicates that she is now headed east. The boat speed is still 30 mph. The wind speed is still a steady 10 mph from the southwest.

---

[1] Use program VECTOR.2D in the appropriate appendix to illustrate the addition and subtraction of vectors.

[2] Use program VECTOR.2D in the appropriate appendix to add and subtract vectors, to find dot products, and to find norms.

a. As a result of both the wind and the boat speed, in which direction would you now expect the boat to travel?

b. Let *d* and *w* denote boat and wind speeds and directions respectively. Determine *u* = *d* + *w*.

c. What is the magnitude of vector *u*?

d. In what direction is the boat headed? At what speed?

e. What does vector *u* represent geometrically in terms of vectors *d* and *w*?

f. Notice that vector *b* from part (2) is in the opposite direction of vector *d* from part (3). Compare the results from parts (2g) and (3e).

4. Suppose the speed of the boat is reduced to 15 mph.

    a. If the boat is headed west, what is the actual direction and speed of the boat if the wind is still steady and from the southwest at 10 mph?

    b. In terms of the vectors *b* and *w* in part (2), how can the situation in part (4a) be represented?

    c. If the boat is headed east, what is the actual direction and speed of the boat (same wind conditions)?

    d. In terms of the vectors *d* and *w* in part (3), how can the situation in part (4c) be represented?

5. Complete the following for vectors *a* and *b* shown in the figure at right. Do not assume any particular scales for the *x* and *y* axes.

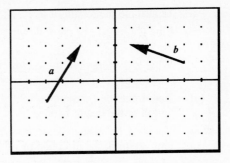

    a. Sketch the vector *a* + *b*.

    b. Sketch the vector *a* – *b*.

    c. State the geometric significance of the sum of two vectors in terms of these vectors.

    d. State the geometric significance of the difference of two vectors in terms of these vectors.

25.4 Algebra of Vectors

Vectors have an algebraic system that is similar to that of the real numbers in some ways, but quite different in others. In Section 25.2 it was discovered that vector addition and scalar multiplication have geometric meanings different from the addition and multiplication of real numbers. The activities that follow allow further exploration into the similarities and differences in the algebraic structures of these systems.

Properties of Vector Addition and Scalar Multiplication

1. a. Let $u = <3, -1>$ and $v = <1, 2>$, $w = <-2, 3>$. Sketch each vector u, v, and w as well as the vector that is the result of the vector arithmetic for the following.

    i. $u + v$ ii. $v + u$ iii. $u + (v + w)$
    iv. $(u + v) + w$ v. $2(u + v)$ vi. $(2 + 3)u$

    b. Is vector addition commutative? Use the definition of vector addition to prove your conjecture for any two vectors $u = <x_1, y_1>$ and $v = <x_2, y_2>$.

    c. Is vector addition associative? Use the definition of vector addition to prove your conjecture for any two vectors $u = <x_1, y_1>$ and $v = <x_2, y_2>$.

    d. Prove that scalar multiplication is distributive over vector addition.

    e. Prove that $(r + s)u = ru + su$ for scalars r and s and vector $u = <x_1, y_1>$.

    f. Determine the identity element for vector addition.

    g. Determine the identity element for scalar multiplication.

    h. When are two vectors equal?

2. Unit vectors are any vectors of length one. Two examples of unit vectors are i and j. It is possible to determine a unit vector in the direction of any vector. Let $u = <3, -1>$.

    a. Determine the length of u.

    b. Find the lengths of the following vectors:

    i. $2u$ ii. $-2u$ iii. $0.3u$ iv. $-0.3u$

    c. What is the length of the vector cv for any scalar c and any vector $v = <a, b>$?

    d. Determine the components of a vector in the direction of u where the length of the vector is one.

    e. If $v = <a, b>$ is any vector, how might you determine the components of a unit vector in the direction of v?

3. How does the length of the sum of two vectors $|u + v|$ compare to the sum of the lengths of the vectors $|u| + |v|$? Explain.

Dot Product

The following definitions are important as meaning for vector algebra continues to develop.

Definition: The **dot product** of the vectors $u = <x_1, y_1>$ and $v = <x_2, y_2>$ is the number $u \cdot v = x_1x_2 + y_1y_2$.

Definition: The **angle between the vectors** u and v is the smaller of the two angles formed between these vectors.

The following activities are designed to help you explore the structure of vector algebra imposed by the dot product. A geometric interpretation for the dot product is also suggested.

1. Complete the following for $u = <3, -1>$ and $v = <1, 2>$, $w = <-2, 3>$, then repeat for the general vectors $u = <x_1, y_1>$, $v = <x_2, y_2>$, and $w = <x_3, y_3>$.

   a. Determine the value of $|u|$. Determine $u \cdot u$. How do these values compare?

   b. Is the dot product commutative?

   c. Is the dot product associative?

   d. Determine the value of $2(u \cdot v)$. Can this value be obtained if 2 is a scalar factor of u or of v? Explain.

2. a. Find a vector v that is perpendicular to the vector $u = <3, -1>$. Explain your work.

   b. What is the measure of the angle between u and v?

   c. What is the dot product $u \cdot v$?

   d. Determine the dot product of the unit coordinate vectors i and j.

   e. What is the measure of the angle between i and j ?

   f. The dot product is helpful in determining the measure of the angle between two vectors. What would you expect the dot product of two perpendicular vectors to be? Why?

3. Consider vectors $v = <a, b>$ and $w = <c, d>$ in the figure. Let θ be the angle between v and w. Let $\beta = \theta + \alpha$.

   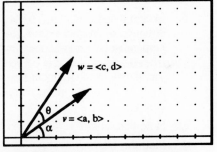

   a. Use the angle addition formulas for sine and cosine to determine an expression for $\cos \theta = \cos (\beta - \alpha)$ in terms of $\cos \beta$, $\cos \alpha$, $\sin \beta$, and $\sin \alpha$.

   b. In terms of the components of $v = <a, b>$ and $w = <c, d>$, determine values for the cosines and sines of α and β.

   c. Rewrite the expression found in (3a) with the values found in (3b).

   d. Write the expression found in part (3c) in terms of vector norms and dot products.

   $$\cos \quad \theta \quad = \quad \underline{\hspace{4cm}}$$

   e. Provide a geometric interpretation for the dot product.

4. Consider the boater problem from Section 25.2. Use the dot product to determine the direction angle at which the boat is headed due to wind and boat vectors

   a. for problem (2)   b. for problem (3)

25.5 Components and Projections

The cable car problem discussed on page 144 requires a slightly different analysis than the boat problem. To determine the amount of force required to keep the cable car from rolling downhill, take the force acting on the cable car and break it down into its individual components.

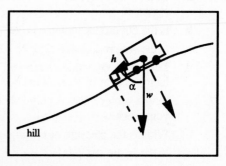

The vector representing the weight of a fully loaded cable car due to its mass and gravity is approximately 9 tons exerted in the $-j$ direction. It is denoted in the figure by the vector w. The cable car must travel in the direction of the downward slope of the hill. From the weight vector, it is possible to determine the component of the weight vector in the direction of the hill, denoted by the vector h. To determine what force must be exerted to keep the car stationary, it is only necessary to determine vector h and exert a force in the opposite direction.

1. Suppose the hill has a slope of 20°. The length of h is the component of w in the direction of the hill. To find h, complete the following.

    a. Find a unit vector u in the direction of the hill.

    b. Sketch u and extend it. To determine the component of a vector w in the direction of another vector, say u, w can be thought of as the sum of two perpendicular vectors, one of which is in the direction of u. Sketch a line from the tip of w that is perpendicular to a vector in the direction of u. The length of the segment starting at the cable car and ending at the right angle is the length of the component of w in the direction of u, denoted $\text{comp}_u w$.

    c. Note that $\cos \alpha = |h|/|w|$ so the magnitude of h is $|w| \cos \alpha$. Find $|h|$.

    d. Find the vector h.

    e. Note that the vector h is the force exerted on the cable car causing it to travel downhill. Determine the force that must be exerted to keep it stationary.

2. a. Summarize the above in a single equation, using the equation for the cosine of the angle between two vectors found in Section 25.4.

    $$|h| = \text{comp}_u w = |w| \cos \alpha = \underline{\qquad\qquad}$$

    b. To determine the vector w in the direction of u, called the projection of w in the direction of u and denoted $\text{proj}_u w$, multiply the scalar $\text{comp}_u w = |h|$ by the unit vector in the same direction as h.

    $$\text{proj}_u w = \text{comp}_u w \, \frac{u}{|u|} = \underline{\qquad\qquad}$$

A

CASIO Graphics Calculator Appendix

A.1 CASIO Programs for *Exploring Calculus*

Listed in the first column are the titles of the explorations in this collection. The second column lists the programs necessary to complete each of the explorations. The third column lists the page on which the necessary program can be found.

Exploration	Program(s)	Page
1. Exploring Families of Functions	FAMILY	176
2. Extended Families of Functions	FAMILY (optional)	176
3. Understanding Function Notation	None	
4. Behavior at a Point and End Behavior	BEHAVIOR	170
5. Continuity		
6. Understanding the Derivative	SPIDER	203
	SECANT	199
7. Hierarchy of Functions	None	
8. Graphical Differentiation	TANGENT	206
9. Max/Min Problems	None	
10. Linear Approximations	LINAPPRX (In footnote)	52
	NEWTON-G	182
	NEWTON	
11. Interpreting the Second Derivative	SPIDER	203
	TANGENT	206
12. A Development of the Function $F(x) = e^x$	SECANT or	199
	BEHAVIOR	170
13. Determining Distance from Velocity	AREA	167
	RIEMANN	196

A.2 Using the CASIO Graphics Calculator with *Exploring Calculus*

To make best use of these materials, the following suggestions might be helpful. Throughout the *Exploring Calculus* footnotes and the CASIO Appendix, "CASIO 7000" refers to the calculator models fx-7000G, fx-7000GA, fx-7500G, fx-8000G, and fx-8500G. "CASIO 7700" refers to one model, the CASIO fx-7700G. Programming and operation vary slightly between the 7000 series and the 7700. Specific directions for use of each are provided when the operation of these machines is different. This section describes conventions and error prevention common to CASIO models listed above. Specific directions for the CASIO 7000 series and the CASIO 7700 are provided in Sections A.3 and A.4, respectively.

CONVENTIONS

- To graph functions such as $y = |x|$ and $y = \sqrt{4 - x^2}$ (semi-circle) so that these appear on the screen as expected, use the default viewing rectangle $[-4.7, 4.7]$ by $[-3.1, 3.1]$, or a multiple of this viewing rectangle. To obtain this rectangle, press $\boxed{\text{Range}}$ to view the range screen.

  CASIO 7000   Press $\boxed{\text{SHIFT}}$ followed by $\boxed{\text{DEL}}$ to set the viewing rectangle.

  CASIO 7700   Press $\boxed{\text{F1}}$ to initialize the range to the default view.

  The screen will appear similarly to those at the top of page 153.

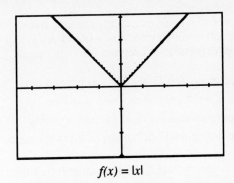

$f(x) = |x|$

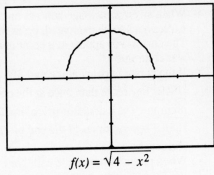

$f(x) = \sqrt{4 - x^2}$

- In the calculus explorations, when the suggested viewing rectangle is given as [−5, 5] by [−3, 3], the default viewing rectangle is preferred.
- In the Range, the scale settings determine the tick marks on the *x*- and *y*-axes. Scale markings are not suggested in the materials. Choosing a scale that will allow about ten tick marks to be plotted will help you to interpret graphs.
- To interrupt a graph as it is being drawn, press $\boxed{\text{AC}}$.
- To terminate a program, press $\boxed{\text{AC}}$ or $\boxed{\text{MODE}}$ 1.
- The cursor keys operate similarly to cursor control keys on a computer. In a program, these can be used to move the cursor to any position in the text to edit the program.
- The insert and delete keys operate similarly to those on a computer. Pressing $\boxed{\text{SHIFT}}$ $\boxed{\text{INS}}$ allows characters to be inserted in the location immediately preceding the highlighted position. Pressing $\boxed{\text{DEL}}$ when a character is highlighted will delete the highlighted character.
- Keep program 2 empty until it is needed to store program Cls-TEXT. Cls-TEXT is used as a subroutine in programs that plot points. This program is found in Section A.8 of this appendix. Users of fx-7000GA calculators do not need Cls-TEXT.

PREVENTING OR CORRECTING COMMON PROBLEMS

- Check the mode in the calculator whenever an unexpected answer or graph appears. The calculator should be set to **radian** mode for *Exploring Calculus*. Graphs of the trigonometric functions do not appear on the screen for the viewing rectangles suggested if the calculator is set for degree mode. To put the calculator into radian mode

CASIO 7000   Press $\boxed{\text{MODE}}$ $\boxed{5}$ $\boxed{\text{EXE}}$

CASIO 7700   Press $\boxed{\text{SHIFT}}$ $\boxed{1}$ (for DRG menu) then $\boxed{\text{F2}}$ for radians

- When an error message appears on the screen while executing programs and expressions to be evaluated, press the left or right cursor arrow (REPLAY). The cursor will appear at a point in the expression or program where the error was committed.
- If the cursor appears at the end of a program, it is likely that you pressed the EXE key more than once at the end of a program when you entered the program. The calculator is looking for more commands to execute. Pressing DEL several times at the end of the program will delete any of these invisible characters.
- When programming code calls for E–6, use the \x(EXP) key to enter E (use the (–) key to enter the negative sign on CASIO 7000).
- In programs, all characters that appear as 0 are the number zero. The letter O (oh) was not used as a variable in any of the programs.
- CAUTION: Take care when putting the calculator in MODE 3, PCL. This is the mode in which all program memory in the calculator can be inadvertently erased. To clear a single program, press MODE 3, move the cursor to the location indicating the program to be deleted, press AC, then immediately press MODE 1 to avoid clearing any other (or all) programs from the calculator memory.

A.3 Using the CASIO 7000 with *Exploring Calculus*

CONVENTIONS

- Store expressions to be evaluated several times in program 0. Programs that evaluate functions use program 0 as a subroutine for function evaluation.
- Store graph statements for functions to be investigated in program 1 to avoid having to key them in more than once. Programs that graph functions have all been written to use program 1 as a subroutine for the graph statements.
- Program locations 0, 1, 2, and 3 are used as subroutines in other programs as follows. Do not store other programs in these locations.

Prog 0   Function evaluation
      1   Graph statement
      2   Program Cls-TEXT
      3   NDERIV (when needed for TANGENT and NEWTON)

- CASIO fx-7000G and fx-7000GA models have very limited memories. Users will find that the program memory fills with only three or four programs stored. To alleviate some of the difficulty with removing and restoring programs, make programs as short as possible. Possibilities include shortening program titles to some meaningful acronym and shortening messages printed on the screen to identify output. A few of the programs in this appendix include adaptations for an abbreviated form.

PREVENTING OR CORRECTING COMMON PROBLEMS

- For the CASIO 7000, certain functions cannot be accessed when the calculator is in any of the statistical modes or in base-n mode. If the calculator will not allow access to the absolute value or root functions, it is probably in one of the statistical modes.

CASIO 7000 REFERENCE SHEET

Option	Basic Keys	To do:
Clear	AC	Clears the text screen.
Execute (=)	EXE	Executes a command; works like the = key on a conventional calculator.
Shift	SHIFT	Press to access options printed in yellow above the keys.
Alpha characters	ALPHA	Press to access options printed in red below the keys.
Graph to Text	G↔T	Toggles between graphics and text screens.
Replay	Replay	The left and right cursor arrow keys; allows last executed line to be edited and re-executed.
negative numbers	(−)	Negative sign (use to indicate negative numbers).
Answer	Ans	Returns the value of the last executed computation.
Operating mode	M Disp	Holding down key displays current operating mode of calculator.
Mode Settings	MODE \<key\>	Puts calculator into the mode indicated on the template listed below the display.
Program n	Prog n	Calls program in memory location n.
π	SHIFT EXP	Gives stored estimate of π.
Insert	SHIFT ⇒	Allows text to be inserted before the character highlighted by the cursor.

Graphics Commands

Option	Basic Operations	To do:
Graph	[Graph] <function in x>	Graphs function following [Graph] command.
Cls	[SHIFT] [G↔T]	Clears the graphics screen.
Range	[Range]	Allows user to set the domain and range of the viewing rectangle.
Default Viewing Rectangle	[Range] [SHIFT] [DEL]	Sets Range to default values of [–4.7, 4.7] by [–3.1, 3.1].
Trace	[SHIFT] [Graph]	Allows user to determine x (and y) coordinates of points along an active graph.
X↔Y	[SHIFT] [⇓]	Toggles between x- and y-coordinates of points on graphics screen when using trace or plot options.
Zoom-in	[SHIFT] [×]	Zoom in by a factor of 2.
Zoom-out	[SHIFT] [÷]	Zoom out by a factor of 2.
Factor c	[SHIFT] [Range] c	Multiplies current Range values by a factor of $\frac{1}{c}$ for some nonzero number c.
Factor c, d	[SHIFT] [Range] c, d	Multiplies current x-Range values by a factor of $\frac{1}{c}$, and current y-Range values by $\frac{1}{d}$, for some nonzero numbers, c and d.
Plot m, n	[SHIFT] [M] Disp m, n	Plots the point (m, n) where m and n are two numbers in the current calculator Range.

Arrow Keys

n [→] A	Store n in memory location A.
[SHIFT] [7]	Logical arrow — used in programming.
[⇒]	Right cursor arrow.

A.4 Using the CASIO 7700 with *Exploring Calculus*

CONVENTIONS

- Enter expressions to be evaluated several times or to be graphed more than once into the function menu ($\boxed{\text{SHIFT}}$ $\boxed{0}$) to avoid having to key them in more than once.
- Programs that evaluate or graph functions have all been written to use functions stored in f_1. Some programs also use function f_2.
- Store program Cls-TEXT in program 2. It is used as a subroutine in programs that plot functions by points and is found in Section A.8 on page 173.

PREVENTING OR CORRECTING COMMON PROBLEMS

- If vertical lines appear when you are graphing functions with vertical asymptotes, change the graphing mode to point plotting and redraw the graph.

CASIO 7700 REFERENCE SHEET I

Option	Basic Keys	To do:
X, θ, T	$\boxed{\text{X,}\theta\text{,T}}$	Displays appropriate variable for graphing mode of calculator.
Clear	$\boxed{\text{AC}}$	Clears the text screen.
Execute (=)	$\boxed{\text{EXE}}$	Executes a command; works like the $\boxed{=}$ key on a conventional calculator.
Shift	$\boxed{\text{SHIFT}}$	Press to access options printed in yellow or menus printed in green above the keys.
Alpha characters	$\boxed{\text{ALPHA}}$	Press to access options printed in red above the keys.
Graph to Text	$\boxed{\text{G}\leftrightarrow\text{T}}$	Toggles between graphics and text screens.
Replay	Replay	The left and right cursor arrow keys; allows last executed line to be edited and re-executed.
Answer	$\boxed{\text{Ans}}$	Returns the value of the last executed computation.

Option	Basic Keys	To do:
Calculation or system mode	MODE	Allows user to set computation and system modes.
graphics or statistical mode	MODE SHIFT	Allows user to set the graphics and statistical modes of the calculator.
Program n	SHIFT Range F3 n	Calls program in memory location n.
π	SHIFT EXP	Gives stored estimate of π.
Insert	SHIFT DEL	Allows text to be inserted before the character highlighted by the cursor.

Graphics Commands

Option	Basic Operations	To do:
Graph	Graph	Graphs function following [Graph] command. Function is in terms of X, θ, or T depending on graphics mode.
Range	Range	Allows user to set the domain and range of the viewing rectangle.
Default Viewing Rectangle	Range F1	Sets Range to default values of [–4.7, 4.7] by [–3.1, 3.1].
Trace	F1	Allows user to determine x- (and y-) coordinates of points along an active graph.
Zoom in	F2 F3	Zoom in by a factor set in calculator.
Zoom out	F2 F3	Zoom out by a factor set in calculator.
Plot	F3	Plots a point at the center of the graphics screen.
Cls	F5	Clears the graphics screen.
Display coordinates	F6	Changes display of x- and y-coordinates of points on graphics screen when using trace or plot options.

Arrow Keys

n → A	Store n in memory location A.
* SHIFT 7	Logical arrow — used in programming.

CASIO 7700 REFERENCE SHEET II

The CASIO 7700 graphics calculator has been developed to operate from menus. This reference sheet was created to help you locate the commands and operations that are found in the calculator menus. The calculator keys are listed in boxes. Menus accessed from the calculator keys are typed in capital letters. Only menus used in the calculus materials are included here.

Mode Screens

The screen on the right shows the available system and calculation modes of the CASIO 7700. These are displayed by pressing $\boxed{\text{MODE}}$. The calculator will most often operate in the 1: RUN mode. System modes 2 and 3 are used to write or edit programs (2: WRT) and delete programs (3: PCL). The CASIO 7700 allows normal calculations of rational numbers (+: COMP),

Sys mode	Cal mode
1: RUN	+: COMP
2: WRT	−: BASE-N
3: PCL	×: SD
REG model	+: REG
4: LIN	∅: MATRIX
5: LOG	**Contrast**
6: EXP	←: LIGHT
7: PWR	→: DARK

calcuations in various number bases (−: BASE-N), standard deviation, two-variable statistical computations (×: SD), regression, one-variable statistical computations (+: REG), and operations with matrices (∅: MATRIX). To change the system or calculation mode, press the number preceding the desired mode.

The graph and statistical modes are displayed on the right. These are accessed by pressing $\boxed{\text{MODE}}$ followed by $\boxed{\text{SHIFT}}$. The CASIO 7700 has built-in capabilities to graph rectangular (+: REC), polar (−: POL) and parametric (×: PARAM) equations and rectangular inequalities (+: INEQ). Graphs can be plotted by points (6: PLOT) or connected (5: CONNECT). The statistical modes are not used in *Exploring Calculus*.

Stat data	Graph type
1: STO	+: REC
2: NON-	−: POL
Stat graph	×: PARAM
3: DRAW	+: INEQ
4: NON-	**Draw type**
	5: CONNECT
	6: PLOT

Menus and Graphing Utilities

The CASIO 7700 has been streamlined to use menus for graphing utilities, many mathematical functions, programming commands, angle modes, display modes, memory clearing, and function storage. The graphing options/menus are accessed by directly pressing the function (F) keys $\boxed{\text{F1}}$ through $\boxed{\text{F6}}$. The menus are accessed by pressing $\boxed{\text{SHIFT}}$ then pressing the key with the name of the menu written above it in green.

Graphing Utilities

The graphing utilities are written above the green function keys as shown. Press the corresponding F key. While a

Trace Zoom Plot Line Cls Coord
 (F1)   (F2) (F3) (F4) (F5) (F6)

graph is displayed on the screen, it may be traced ($\boxed{\text{F1}}$) with corresponding x and y coordinates displayed. More accuracy for x and y may be obtained by pressing $\boxed{\text{F6}}$. To plot a point on the screen without tracing along the graph, press $\boxed{\text{F3}}$. To clear the graphics screen, press $\boxed{\text{F5}}$ for Cls.

The Zoom ($\boxed{\text{F2}}$) menu allows the graph to be viewed from different viewing rectangles. Zoom BOX ($\boxed{\text{F1}}$) allows the user to zoom in by drawing a box around the part of the graph to be magnified. It is also possible to zoom in (xf, $\boxed{\text{F3}}$) or out ($x^{1/f}$, $\boxed{\text{F4}}$) by a factor set in the ZOOM menu (FCT, $\boxed{\text{F2}}$). The original viewing rectangle is saved and can be recalled (ORG, $\boxed{\text{F5}}$).

When you press $\boxed{\text{SHIFT}}$, the graphing utilities are displayed on the screen. The Plot, Line, and Clear Screen (Cls) options operate as described above. However, a graph does not have to be displayed on the screen to use these options.

Menus

The menus for the CASIO are listed below, with sub-menus and sub-submenus indicated. The F key(s) to be pressed are also indicated. To access a menu, press $\boxed{\text{SHIFT}}$ followed by the key over which the menu title is listed in green. Choose the function key corresponding to the option to be used. In some cases, sub-menus exist. Some of these sub-menus also have sub-menus. To return to the previous menu, press $\boxed{\text{PRE}}$. Pressing $\boxed{\text{PRE}}$ more than a few times results in a screen with no menu listed. Examples for use of the menus follow the menu lists.

MATH

	(F1)	(F2)	(F3)	(F4)
	HYP	PRB	NUM	DMS
(F1)	snh	X !	Abs	° ′ ″
(F2)	csh	nPr	Int	° ′″
(F3)	tnh	nCr	Frc	
(F4)	snh^{-1}	Rn#	Rnd	
(F5)	csh^{-1}			
(F6)	tnh^{-1}			

Example: To access the absolute value function, press $\boxed{\text{SHIFT}}$ then MATH (written above the $\boxed{\text{GRAPH}}$ key), $\boxed{\text{F3}}$ for the NUM sub-menu, then $\boxed{\text{F1}}$ for Abs (absolute value).

PRGM

(F1)	(F2)	(F3)	(F4)	(F5)	(F6)
JMP	REL	Prg	?	◢	:

(F1)	⇨	=
(F2)	Goto	≠
(F3)	Lbl	>
(F4)	Dsz	<
(F5)	Isz	≥
(F6)		≤

The programming commands are listed in the PRGM menu. The colon and question mark can also be used for programming in the interactive mode by first accessing this menu. All of the equivalence relations are found in the REL sub-menu. The jump commands are found in the JMP menu. To run a program, access the PRGM menu by pressing $\boxed{\text{SHIFT}}$ then PRGM (written above the $\boxed{\text{Range}}$ key). Press $\boxed{\text{F3}}$ for Prg. Prog will appear on the screen. Press the number or letter that is the location of the desired program. Press $\boxed{\text{EXE}}$ to run the program.

DRG

(F1)	(F2)	(F3)	(F4)	(F5)	(F6)
Deg	Rad	Gra	°	r	g

The DRG menu is used to set the calculator angle mode to degrees (Deg), radians (Rad) or gradians (Gra). The other symbols are used to convert one number at a time.

DISP

(F1)	(F2)	(F3)	(F4)
Fix	Sci	Nrm	Eng

The DISP (display) menu is used display results of computations in fixed (Fix), scientific (Sci), Floating point (Norm), or Engineering (Eng) notation.

CLR

	(F1)	(F2)	(F3)	(F4)
	Mcl	Scl	ARR	PRG
(F1)			YES	YES
(F6)			NO	NO

The CLR menu is used to clear the values stored in memory locations A through Z, r and θ (Mcl), the statistical memories (Scl), the array memory (ARR), and all of the program memory (PRG). A second chance is provided when you clear array and program memories. These options are not reversable and should be used carefully.

$\boxed{\text{F}}$ MEM

$\boxed{\text{F1}}$	$\boxed{\text{F2}}$	$\boxed{\text{F3}}$	$\boxed{\text{F4}}$
STO	RCL	fn	LIST

The CASIO 7700 allows functions to be stored for use in function evaluation and graphing. For N = 1, 2, 3, ... , 6 STO N stores a function listed on the screen in function location N. RCL N prints a copy of the function stored in function location N on the screen. The fn option lists a reference to function f_N on the screen. This is helpful when you are using the same function several times in programs, for graphing, or for function evaluation. LIST shows the first 14 characters of each function f_1 through f_6.

To store the function $y = x^2 - 3x + 4$ in function f_1, enter $X^2 - 3X + 4$, access the function memory ($\boxed{\text{F}}$ MEM) and press $\boxed{\text{F1}}$ 1 then $\boxed{\text{EXE}}$. Press $\boxed{\text{F4}}$ make sure that the function is listed in f_1.

It is possible to use the function memory to store programs. It is a useful copy-and-paste tool in the following way. Store a program is the program memory. While in the WRT (write) mode, access the function memory ($\boxed{\text{F}}$ MEM) and press $\boxed{\text{F1}}$ 1 then $\boxed{\text{EXE}}$. This will store the program in function f_1. Exit the program ($\boxed{\text{MODE}}$ 1). Open an empty program location, access the function memory ($\boxed{\text{F}}$ MEM), and press $\boxed{\text{F2}}$ 1 to recall function f_1. A working copy of the entire program stored in f_1 is now stored in the open program location. This feature is helpful when you are using programs that have similar programming steps to be modified for a new program. It is also helpful when you are attempting to modify programs.

A.5 Diving Board Problem[1]

When a person stands on a diving board, the amount the board bends at the point she is standing, y, below its rest position is a cubic function of x, the distance from the built-in end to the point where she is standing on the board. See the drawing on page 163.

---

[1] This problem is adapted from Paul Foerster, *Algebra and Trigonometry*, Addison-Wesley, 1990. It is included here to allow students to solve a problem from precalculus while becoming familiar with the CASIO graphics calculator.

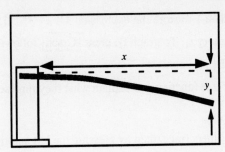

Suppose that the following measurements are taken. Note that the deflection from the horizontal (the bend) is recorded as a negative value.

Horizontal distance in feet from built-in end of board x	Deflection from horizontal in thousandths of an inch y
0	0
2	–528
4	–1784
6	–3576

1. Derive the equation expressing y as a function of x.

    Write $y = ax^3 + bx^2 + cx + d$ and then use the four measured values for x and y to determine a system of four equations in four unknowns (a, b, c, and d). (Set up the matrices and solve this on the CASIO 7700. Otherwise, complete by hand.)

2. What does it mean in this situation for x to be positive? negative? Do these make sense? Explain.

3. What does it mean in this situation for y to be positive? negative? Do these make sense? Explain.

4. What is an appropriate domain for x? What is an appropriate range? Explain.

    To graph the function on the suggested domain and range, enter these values into the calculator. Press the $\boxed{\text{Range}}$ key and enter the appropriate values, following each with $\boxed{\text{EXE}}$.

5. Sketch the graph of this function.
    CASIO 7000
    Press $\boxed{\text{Graph}}$ 4 $\boxed{\text{ALPHA}}$ $\boxed{\text{X}}$ $\boxed{x^y}$ 3 – 115 $\boxed{\text{ALPHA}}$ $\boxed{\text{X}}$ $\boxed{x^2}$
    – 50 $\boxed{\text{ALPHA}}$ $\boxed{\text{X}}$ $\boxed{\text{EXE}}$.

    CASIO 7700
    Enter the formula for the function into function location f_1 as follows:
    Press $\boxed{\text{SHIFT}}$ then 0 for $\boxed{\text{F}}$ MEM. Enter the equation for the function:

    4 $\boxed{\text{X,θ,T}}$ $\boxed{x^y}$ 3 – 115 $\boxed{\text{X,θ,T}}$ $\boxed{\sqrt{}}$ (for x^2) – 50 $\boxed{\text{X,θ,T}}$

Press $\boxed{\text{F1}}$ then 1 to store the expression $4X \, x^y \, 3 - 115X^2 - 50X$ in function memory f_1. To graph f_1, press $\boxed{\text{Graph}}$ followed by $\boxed{\text{F3}}$ 1 (for f_1), the $\boxed{\text{EXE}}$.

6. Find the zeros of this function and explain what they represent in this situation.

CASIO 7000

Press $\boxed{\text{Shift}}$ then Trace, then by pressing the left and right cursor keys move the blinking cursor to a point where the graph crosses the x-axis. The x-coordinate of the blinking cursor will be displayed in the lower left corner of the screen. To view the corresponding y-coordinate, press $\boxed{\text{Shift}}$ $\boxed{\text{X} \leftrightarrow \text{Y}}$. Press $\boxed{\text{Shift}}$ $\boxed{\times}$ to activate the automatic zoom-in.

Press $\boxed{\text{Range}}$ to view the new Range (and to determine the zoom factor).

Press $\boxed{\text{Range}}$ a second time to view the graph again. Trace to the intersection point and zoom in again to determine the coordinates of a better approximation of the coordinates of this zero. Repeat this process of tracing and zooming in until you can approximate the zero of the function to the nearest tenth.

Reset the Range to the values in part (4). Graph the function again and then use Trace and the automatic zoom-in to determine the coordinates of the other zeros.

CASIO 7700

While the graphics screen is visible, press $\boxed{\text{F1}}$ to Trace. Use left and right cursor arrows to move the cross hairs to a point near where the graph crosses the x-axis. Press $\boxed{\text{F2}}$ to zoom in on the function and choose $\boxed{\text{F1}}$ for BOX. Position the cross hairs at one corner of a box to enclose the intersection point. Press $\boxed{\text{EXE}}$. Move the cursor to the opposite diagonal corner. Notice that a box is forming. Enclose the intersection point and press $\boxed{\text{EXE}}$. The new viewing rectangle has the dimensions of the box just sketched.

Press $\boxed{\text{F1}}$ to trace to the intersection point. (If nothing happens, keep pressing the right-arrow key until the cross hairs appear on the graph). Both x and y coordinates are shown. For more accuracy, press $\boxed{\text{F6}}$ to display x, then y, with more significant digits.

7. Suppose that the board is 10 feet long. How far does its tip sag below the horizontal when you stand at the end of the board? Repeat for a board that is 12 feet long.

CASIO 7000

To avoid having to enter the function each time you want to graph it, enter it as a program. Once you have entered the program, graph the desired function by pressing $\boxed{\text{Prog}}$ n $\boxed{\text{EXE}}$ (where n is the number of the location of the program). To program press $\boxed{\text{Mode}}$ 2. Move the blinking cursor over the number 1 and press $\boxed{\text{EXE}}$. Now enter the program into program 1 by pressing $\boxed{\text{Graph}}$ 4 $\boxed{\text{ALPHA}}$ $\boxed{\text{X}}$ $\boxed{x^y}$ 3 – 115 $\boxed{\text{ALPHA}}$ $\boxed{\text{X}}$ $\boxed{x^2}$ – 50 $\boxed{\text{ALPHA}}$ $\boxed{\text{X}}$.

This is the entire program. **Note: Do NOT press** $\boxed{\text{EXE}}$ **at the end of the program.** Press $\boxed{\text{Mode}}$ 1 to enter the program into memory and to return to the RUN mode. Reset the Range to the values in part (4) again and press $\boxed{\text{Prog}}$ 1 to run the program.

CASIO 7700

Graph the function stored in f_1 by pressing $\boxed{\text{Graph}}$ followed by $\boxed{\text{SHIFT}}$ 0 then $\boxed{\text{F3}}$ 1 to print f_1 on the screen. Press $\boxed{\text{EXE}}$ to graph f_1.

Both CASIOs

To answer question (7) graphically:

Use the Trace key to move the blinking cursor to the point whose x-coordinate is as close to 10 as possible.

CASIO 7000: Use the $\boxed{\text{X↔Y}}$ key to find the y-coordinate.

CASIO 7700: Press $\boxed{\text{F1}}$ to trace the curve. Both x and y coordinates are displayed on the screen. Press $\boxed{\text{F6}}$ to display x and y with more accuracy.

Zoom in and repeat the Trace process. Repeat until the x-coordinate is as close as desired to 10 and then determine the corresponding y-coordinate. Will the answer you obtained for y in this manner be less than or greater than the value of y when x is exactly equal to 10? Explain. Repeat this process for the 12-foot board.

To answer question (7) numerically:

CASIO 7000

a. Enter 4 $\boxed{\times}$ 10 $\boxed{x^y}$ 3 – 115 $\boxed{\times}$ 10 $\boxed{x^2}$ – 50 $\boxed{\times}$ 10 $\boxed{\text{EXE}}$.

To find the value for x = 12, press $\boxed{\Leftarrow}$ for replay, replace each 10 with 12, and press $\boxed{\text{EXE}}$. (Notice that the word REPLAY is written above the arrows $\boxed{\Leftarrow}$ and $\boxed{\Rightarrow}$. Pressing either of these arrows allows the last calculation to be *replayed*, also allowing editing of the last expression entered.)

b. Enter the following code:

$\boxed{\text{Shift}}$ $\boxed{?}$ $\boxed{\rightarrow}$ $\boxed{\text{ALPHA}}$ $\boxed{\text{X}}$ $\boxed{:}$ 4 $\boxed{\text{ALPHA}}$ $\boxed{\text{X}}$ $\boxed{x^y}$ 3 – 115

$\boxed{\text{ALPHA}}$ $\boxed{\text{X}}$ $\boxed{x^2}$ $\boxed{-}$ 50 $\boxed{\text{ALPHA}}$ $\boxed{\text{X}}$ $\boxed{\text{EXE}}$.

A question mark appears. Enter 10 $\boxed{\text{EXE}}$ to find the y-coordinate,
when $x = 10$.   Then press $\boxed{\text{EXE}}$ to enter another number. Press 12
$\boxed{\text{EXE}}$ to find the y-value when $x = 12$. This method can be used to
build a table of values for a given function.

CASIO 7700

Store 10 in x by pressing 10 $\boxed{\rightarrow}$ $\boxed{\text{X},\theta,\text{T}}$. Call up function f_1 by
accessing the function memory and pressing $\boxed{\text{F3}}$ 1. Press $\boxed{\text{EXE}}$ to
display the function value. Repeat for $x = 12$.

8.   How far from the built-in end of the board must you stand for the deflection at
that point to be 4 inches? 0.8 inches? Explain.

CASIO 7000

Change program 1 by pressing $\boxed{\text{Mode}}$ 2, moving the blinking cursor over
the place where 1 should be, and pressing $\boxed{\text{EXE}}$. The program is
displayed. With the blinking cursor at the beginning of the first line, press
$\boxed{\text{Shift}}$ $\boxed{\text{INS}}$ (to insert characters). Insert $\boxed{\text{Graph}}$ –4000 $\boxed{\text{EXE}}$. When
the program is run, the horizontal line y = –4000 will be graphed. Why
might graphing this line be helpful in answering question (8)?

Press $\boxed{\text{Mode}}$ 1 to save the program. Reset the Range and run program 1.

Use Trace, automatic zoom-in, and $\boxed{\text{X}\leftrightarrow\text{Y}}$ to determine the x-coordinate
of the point whose y-coordinate is –4000. (Note: The statement for
graphing y = –4000 is placed at the beginning of the program because
when the $\boxed{\text{Trace}}$ key is used, it will trace along the last function graphed.
In what follows, it will be necessary to trace along the graph of
$y = 4x^3 - 115x^2 - 50x$.)

Change program 1 appropriately to answer the second question.

CASIO 7700

Enter –4000 into functin f_2 in the function memory. Enter the graph
statements for the functions as follows: Graph Y = f_2 press
$\boxed{\text{SHIFT}}$ $\boxed{\text{EXE}}$ then enter Graph Y = f_2 $\boxed{\text{EXE}}$. The command $\boxed{\text{SHIFT}}$
$\boxed{\text{EXE}}$ concatenates the two commands so the calculator reads them as
being related. If both functions are graphed as indicated, the CASIO 7700
allows both graphs to be traced. Pressing the up and down cursor keys
moves the trace cross hairs from one graph to the other. Use trace and
zoom several times to determine an accurate estimate for the intersection.

9.   If this function is accurate for diving boards up to 30 feet in length, what
length of board will have the greatest horizontal deflection if you are standing
at the end of it? What will be the horizontal deflection for this board?

CASIO 7000
Reset the Range as before. Edit program 1 by deleting the line of code that graphs the horizontal line. Place the blinking cursor at the start of the line and press the $\boxed{\text{DEL}}$ key (delete) until that line of code is eliminated. Run program 1, then determine the coordinates of the lowest point on the graph.

CASIO 7700
Reset the Range. Graph f_1. Trace along the curve to locate the lowest point on the curve, where y is the smallest. Use $\boxed{\text{F5}}$ (Coord) to display the y-coordinate accurately.

10. Considering the results of part (9), does this function work for boards of lengths up to 30 feet? Explain.

11. How would this situation change if a smaller or larger person than you were to stand on the diving board?

12. What other questions might you ask about this situation? How might you change this situation to make it more interesting to you?

13. What other situations not involving diving boards might have similar mathematical models? Explain.

A.6 Area

Program AREA assists the user in building intuition for the use of rectangles to approximate the area under the curve. The program sketches the graph of a function $y = f(x)$. If the function is monotonic, AREA draws two approximating rectangles, one showing an underestimate for the curve on the screen, the other showing an overestimate for the curve on the screen. The sizes of these areas can be compared visually. Running the program several times, each time reducing the size of the domain without changing the range, allows the user to view approximating rectangles and compare these to the actual area under the curve of the function over successively smaller intervals for x.

After viewing the graphical information from AREA several times, display the values of the areas of the lower and upper approximating rectangles. Additional code is provided for that purpose.

CODE: AREA

The following code provides a graphical experience in determining the usefulness of using rectangles to approximate the area under the curve of a monotonic function.

The graphical code requires 87 (CASIO 7000) or 83 (CASIO 7700) bytes of memory.

Code (CASIO 7000)	Code (CASIO 7700)	Comments
"AREA"	AREA	Title of program.
"XMAX"? → B	"XMAX"? → B	Input XMAX store in B.
Lbl 1	Lbl 1	First Lbl to Goto.
"XMIN"? → A	"XMIN"? → A	Input XMIN store in A.
Range A, B, 1	Range A, B, 1	Set the x-interval.
Prog 1	Graph Y= f_1	Sketch graph of function in Prog 1 or f_1.
Y → R	f_1 → R	Store value of Y or f_1 in R.
Plot B, 0	Plot B, 0	Plot point (B, 0).
Plot B, R	Plot B, R	Plot point (B, R).
Line	Line	Sketch line between points.
A → X	A → X	Store A in X.
Prog 0:Ans → L	f_1 → L	Store value f(x) or f_1 in L.
Plot A, 0	Plot A, 0	Plot point (A, 0).
Plot A, L	Plot A, L	Plot point (A, L).
Line◢	Line◢	Sketch line between points.
Graph Y=L	Graph Y=L	Sketch line $y = L$.
Graph Y=R	Graph Y=R	Sketch line $y = R$.
Goto 1	Goto 1	Return control to Lbl 1.

For a function f that is monotonic, either always increasing or always decreasing, over the domain chosen for the viewing rectangle, add the following code to view the values of the areas of the lower and upper approximating rectangles. The values of the areas are displayed as left (L.AREA) and right (R.AREA) areas. You determine which of these is lower and which is upper. Program AREA also computes and displays the difference between these areas. The following code can be inserted into program AREA immediately above the line **Goto 1**. The full code requires 131 (CASIO 7000) or 127 (CASIO 7700) bytes of memory.

Code (CASIO 7000)	Code (CASIO 7700)	Comments
B–A → H	B–A → H	Determine x-interval width.
"L.AREA":LH◢	"L.AREA":LH◢	Display L.AREA. Determine area of rectangle.
"R.AREA":RH◢	"R.AREA":RH◢	Display R.AREA. Determine area of rectangle.

(continued)

Code (CASIO 7000)	Code (CASIO 7700)	Comments
"DIFF":H(R–L)◢	"DIFF":H(R–L)◢	Determine and display difference left and right area.

OPERATION

Preparation

Set the range to display the graph of the function on an interval over which it is monotonic. Choose the complete range for the domain to be used in program AREA. For illustration purposes, it is best to choose an interval for which $f(\text{Xmax}) \neq 0$.

CASIO 7000

Enter the formula for the function to be graphed into function f_1.

CASIO 7700

Enter the formula for the function to be investigated in program $\emptyset$. Enter the graph statement for the function to be investigated into program 1.

Input

Enter the values for XMAX and XMIN at the appropriate prompts. XMAX is entered first because this value does not change throughout the graphic investigation. Press the $\boxed{\text{EXE}}$ key to continue.

Output

The graph of f is sketched on the interval set by the user, with the values of XMIN and XMAX chosen from the program. Vertical lines are sketched from the x-axis to the graph of f. Press $\boxed{\text{EXE}}$ to sketch the horizontal lines $y = \text{Ymin}$ and $y = \text{Ymax}$. These lines are actually $y = f(\text{Xmin})$ and $y = f(\text{Xmax})$. They correspond with $y = \text{Ymin}$ and $y = \text{Ymax}$ for monotonic functions.

To build intuition for using rectangles to estimate the area under the curve, run AREA several times. Each time use a value of Xmin that is larger than the previous choice, but smaller than Xmax. When the difference between Xmin and Xmax is small enough, the graph of f is imperceptible from the graphs of $y = f(\text{Xmin})$ and $y = f(\text{Xmax})$.

After using the first set of code to investigate under- and overestimates of area using approximating rectangular regions, enter the rest of the code into program AREA. In addition to the output described, the amended program displays the area of the lower approximating rectangle, the upper approximating sum, and the difference in these values.

Sample Run

Preparation

To illustrate the use of AREA, set the range to graph $f(x) = \sin x$ for y in [0, 1.5].

CASIO 7000

Enter the formula $\sin X$ into f_1.

CASIO 7700

Enter program $\emptyset$ as $\sin X$. Enter the graph statement for the $y = \sin x$ into program 1 as Graph $Y = \sin X$.

Run

Run program AREA several times, for XMAX $= \pi \div 2$ and XMIN equal to 0, .5, 1, 1.25, and 1.5. Enter the second set of code above and run AREA again with the same choices of Xmin. Output for Xmin $= .5$ is as follows: L.AREA $= 0.5133671057$, R.AREA $= 1.070796327$, and DIFF $= .5574292211$.

A.7 Behavior

Program BEHAVIOR allows the user to investigate the behavior of a function near a given point or to investigate the end behavior (both left and right) of a given function. The program generates a table of values for the function. This program uses 49 (CASIO 7000) or 51 (CASIO 7700) bytes of memory.

CODE

Behavior Near a Point

Code (CASIO 7000)	Code (CASIO 7700)	Comments
"BHVR"	BEHAVIOR	Title of program.
"A"? $\rightarrow$ A	"A"? $\rightarrow$ A	Input initial value for A.
"H"? $\rightarrow$ H	"H"? $\rightarrow$ H	Input increment.
Lbl 1	Lbl 1	First label to Goto.
"X"	"X"	Display X.
A+H $\rightarrow$ X◢	A+H $\rightarrow$ X◢	Store A + H in X. Display value of X.
"Y"	"f_1"	Display Y or f_1.
Prog $\emptyset$◢	f_1◢	Display value of Y.
H÷2 $\rightarrow$ H	H÷2 $\rightarrow$ H	Divide value of H by 2.
Goto 1	Goto 1	Repeat from Lbl 1.

End Behavior

Change the second to the last line in the code given above from

$$H \div 2 \rightarrow H \quad \text{to} \quad H \times 2 \rightarrow H.$$

OPERATION

Preparation

CASIO 7000

Store the function to be evaluated in program $\emptyset$. For example, if the function is $f(x) = x^2 - 10$, then program $\emptyset$ must be entered as $X^2 - 10$.

CASIO 7700

To use this program, store the formula for the function to be used in f_1 in the $\boxed{F}$ MEM menu. For example, if the function is $f(x) = x^2 - 10$, then function f_1 must be entered as $X^2 - 10$.

Input

The input for this program depends on whether it is being used to investigate the behavior of a function near a given point or to investigate the end behavior of a function.

To investigate the behavior of a function near a point, enter the x-coordinate of that point at the **A?** prompt. Enter the value of an increment to be used at the **H?** prompt. The program evaluates the function for $X = A + H$. It then divides the value of H by 2 and repeats the process. To investigate the behavior of the function for values of x to the right of A, input a small (~ 0.5) positive value for H. Enter a small negative value for H ($- 0.5$) to investigate the behavior of the function for values of x to the left of A.

To investigate the end behavior of a function, the input process is essentially the same. The only difference is that the program multiplies the value of H by 2 (instead of dividing it by 2) before repeating the process. To investigate the right end behavior of a function, input a large (~ 100) positive value of H, and for the left end behavior, input a large (-100) negative value of H. To simplify table building, choose A equal to zero when investigating end behavior.

Output

The program outputs successive values of x and y for the function stored in function program $\emptyset$ (CASIO 7000) or f_1 (CASIO 7700). The first value of x is $A + H$ (A and H are entered by the user). The succeeding values of x depend on how the program is being used. For behavior near a point, H is divided by 2. For end behavior H is multiplied by 2. At each pause in the program, press the $\boxed{\text{EXE}}$ key to go on. The program does not terminate. To terminate the program, press the $\boxed{\text{AC}}$ key.

Sample Run

To investigate the behavior of the function $f(x) = \dfrac{(\sin x)}{x}$ for x just smaller than zero, make sure you enter the second to the last line of the code properly and then enter program $\emptyset$ (CASIO 7000) or function f_1 (CASIO 7700) as

$$\sin X \,/\, X$$

Run BEHAVIOR using A = 0 and H = − 0.5. After each pause in the program, press $\boxed{\text{EXE}}$. A table of values for the function is generated using x-values of −0.5, −0.25, −0.125, −0.0625, etc. Corresponding y-values generated are 0.9588510772, 0.989615837, 0.9973978671, and 0.9993490855 respectively. To terminate the program, press the $\boxed{\text{AC}}$ key.

Adaptation

BEHAVIOR can easily be adapted to print out values for up to six functions.

CASIO 7000

To display y-values for additional functions, remove the line **"Y"**, and the symbol ◢ at the end of the line **Prog Ø** ◢ from program BHVR. Change program $\emptyset$ as follows. Enter the following lines of code into program $\emptyset$:

$$\text{"Y1":f(X)◢}$$

$$\text{"Y2":g(X)◢}$$

where f(X) and g(X) are functions of x. To display additional function values, insert the lines as above, replacing 2 with the function number whose values are to be displayed. For example, to display the values of the functions $y = \dfrac{\sin x}{x}$ and $y = \sin x$, enter program $\emptyset$ as

$$\text{"Y1":sin X + X◢}$$

$$\text{"Y2":sin X◢}$$

Run program BHVR.

CASIO 7700

To display y-values for function f_2 as well as function f_1, insert the following lines of code into program BEHAVIOR immediately following the line $\mathbf{f_1}$◢:

$$\text{"}f_2\text{":}f_2\text{◢}$$

To display additional function values, insert the lines as above, replacing f_2 with the function whose values are to be displayed.

A.8 Clear Text Screen

Program Cls-TEXT clears the text screen. It is intended to be used as a subroutine for any program that graphs a function by plotting points. When points are plotted from a program on the CASIO, the program flips back and forth from the graphics to the text screen, causing a fluttering that makes the graph difficult to see as it is being plotted. Cls-TEXT clears the text screen so that the graphics screen is as prominent as possible during this fluttering. The program requires 16 (CASIO 7000) or 21 (CASIO 7700) bytes of memory.

In programs that use Cls-TEXT as a subroutine, it is referred to as program 2. Store subroutine Cls-TEXT in program location 2, or change the number in the calling program code now listed as **Prog 2** to correspond with the program location of Cls-TEXT.

Note: It is not necessary to use this program on CASIO 7000GA calculators because they do not have the fluttering problem when plotting points from a program. Overhead CASIO 7000s also might not require this program.

CODE: Cls-TEXT

Code (CASIO 7000)	Code (CASIO 7700)	Comments
	Cls-TXT	Program title.
8 → F	7 → F	Store 8 (or 7) in F.
Lbl 1	Lbl 1	First place to Goto.
" "	" "	Print a blank line.
Dsz F	Dsz F	Decrease F by 1, stop when zero.
Goto 1	Goto 1	Repeat from Lbl 1.

Note that the title for the CASIO 7000 series has been omitted. It is not possible to indicate a title without this code appearing on the screen when the program is run. Since the purpose of this program is to completely clear the text screen so that point plotted graphs are more visible, the title has been omitted.

OPERATION

Preparation

Enter the program into program 2. In all of the programs that use Cls-TEXT as a subroutine, the code calling this program is listed as Prog 2. This program requires no input.

Output

Cls-TEXT clears the text screen by printing seven blank lines. The blank text screen allows point plotting to be visible from the graphics screen as the program flips back and forth between graphics and text screens.

Sample Run

Run Cls-TEXT to make sure that the text screen is cleared.

A.9 Conics

Program CONICS sketches the graph of the general quadratic whose equation is

$$Ax^2 + Bxy + Cy^2 + Dx + Ey + F = 0$$

where the parameters A, B, C, D, E, and F are chosen by the user and $C \neq 0$. To assure that circles appear as circles and ellipses appear in the proper dimensions, precede use of this program with use of program SQ.SCREEN, also available in this appendix. CONICS requires 130 (CASIO 7000) or 129 (CASIO 7700) bytes of memory.

CODE: CONICS

Code (CASIO 7000)	Code (CASIO 7700)	Comments
"CONICS"	CONICS	Title of program.
"A"? → A	"A"? → A	Input A, store in A.
"B"? → B	"B"? → B	Input B, store in B.
"C"? → C	"C"? → C	Input C, store in C.
"D"? → D	"D"? → D	Input D, store in D.
"E"? → E	"E"? → E	Input E, store in E.
"F"? → F	"F"? → F	Input F, store in F.

(continued)

Code (CASIO 7000)	Code (CASIO 7700)	Comments
Graph Y = (–(BX + E) + $\sqrt{}$ ((BX + E)^2 – 4C (AX^2 + DX + F))) ÷ (2C)	Graph Y = (–(BX + E) + $\sqrt{}$ ((BX + E)^2 – 4C (AX^2 + DX + F))) ÷ (2C)	Graph the other half of the conic section.
Graph Y = (–(BX + E) – $\sqrt{}$ ((BX + E)^2 – 4C (AX^2 + DX + F))) ÷ (2C)	Graph Y = (–(BX + E) – $\sqrt{}$ ((BX + E)^2 – 4C (AX^2 + DX + F))) ÷ (2C)	Graph the other half of the conic section.

OPERATION

Preparation

Write the equation of the conic to be sketched in its general quadratic form as shown on page 174, and determine the values of the parameters A, B, C, D, E, and F. Set the Range to graph the conic section on a suitable viewing rectangle. If a viewing rectangle that uses the same x and y scale is desirable, first run program SQ.SCREEN (page 205). Note: If C = 0, determine the equation of the general quadratic as a function of y and graph it without using this program.

Input

CONICS asks for the values of A, B, C ≠ 0, D, E, and F in the general quadratic equation $Ax^2 + Bxy + Cy^2 + Dx + Ey + F = 0$. Enter these values at the appropriate prompts.

Output

CONICS displays the graph of the equation entered in its general quadratic form. It first displays one half of the conic, and then displays the other half. Each half is a function of y in terms of x.

Sample Run

To graph the ellipse $\frac{x^2}{4} + \frac{y^2}{9} = 1$, set the Range to sketch the ellipse on the default viewing rectangle. Run CONICS. At the appropriate prompt, enter the values of A = .25, B = 0, C = 1 ÷ 9, D = 0, E = 0, and F = –1. CONICS will display the graph of the ellipse.

A.10 Family

The program FAMILY sketches the graphs of members of a family of functions. Family members sketched are related to the function $y = f(x)$ by the parameters A, B, C, and D in the equation

$$y = A \cdot f(Bx + C) + D$$

where f is a function and A, B, C, and D are real numbers. This program uses 37 (CASIO 7000) or 38 (CASIO 7700) bytes of memory.

CODE: FAMILY

Code (CASIO 7000)	Code (CASIO 7700)	Comments
"FMLY"	FAMILY	Title of program.
"A"? → A	"A"? → A	Input A.
"B"? → B	"B"? → B	Input B.
"C"? → C	"C"? → C	Input C.
"D"? → D	"D"? → D	Input D.
Prog 1	Graph Y=f$_1$	Draw graph of $y = A \cdot f(BX + C) + D$.

OPERATION

Preparation

Set the Range to display the family of function whose graphs are to be sketched.

CASIO 7000

Enter the graph statement for the family of functions to be graphed into program 1. For example, to sketch various members of the family of functions defined by $f(x) = A \cdot \sin(Bx + C) + D$, enter program 1 as

$$\text{Graph Y} = \text{Asin} (BX + C) + D.$$

CASIO 7700

Enter the formula for the family of functions to be graphed into function f_1. For example, to sketch various members of the family of functions defined by $f(x) = A \cdot \sin(Bx + C) + D$, enter function f_1 as

$$\text{Asin} (BX + C) + D.$$

Input

When you run this program, input values for the parameters A, B, C, and D at each corresponding prompt.

Note: This program can be used for families of functions defined by fewer than four parameters. For example, if the family of functions is defined by an equation containing two parameters, store the graph statement for the formula in program 1 (CASIO 7000) or the formula in function f_1 (CASIO 7700). Use A and B for the parameters and run FAMILY. The user is asked to input values for C and D (as well as for A and B). In this case, it does not matter what values are entered for C and D because these are not used in the formula for the function.

Output

After the values for A, B, C, and D are input, the program sketches the graph of the member of the function family defined by the values of the entered parameters. After the graph is drawn, the program stops. Press the $\boxed{\text{EXE}}$ key to repeat the program. By repeating the program, graphs of several members of the family of functions can be graphed on the same screen.

Sample Run

Preparation

To use the program FAMILY to sketch the graphs of members of the family of functions defined by $f(x) = A \cdot \sin(Bx + C) + D$, set the Range to [–5, 5] by [–5, 5] with a scale factor of 1 along each axis.

CASIO 7000

Enter program 1 as

$$\text{Graph Y} = \text{Asin (BX + C) + D,}$$

CASIO 7700

Enter function f_1 as

$$\text{Asin (BX + C) + D}$$

Run

Run FAMILY using A = 1, B = 1, C = 0, and D = 0. A sketch of the graph of $y = \sin x$ is displayed. Press the $\boxed{\text{EXE}}$ key to run the program again. Overlay the graph of $y = 3\sin x$ by entering A = 3, B = 1, C = 0, and D = 0. Sketch the graph of $y = 3 \cdot \sin(x + 2) + 1$ by pressing $\boxed{\text{EXE}}$ to begin the program again and using A = 3, B = 1, C = 2, and D = 1. Terminate the program by pressing $\boxed{\text{AC}}$ if a graph is on the screen, or $\boxed{\text{MODE}}$ 1 if the program is prompting for input.

A.11 Integral

The program INTEGRAL outputs various approximations for the definite integral
of a function f over the interval [a, b]. It lists the midpoint approximation of a
Riemann sum with n subintervals, the trapezoid approximation using n
subintervals, and Simpson's approximation for the definite integral using 2n
subintervals. The value of n is supplied by the user. This program uses 176
(CASIO 7000) or 165 (CASIO 7700) bytes of memory.

Note: The built-in numerical integration option on the CASIO 7700 uses
Simpson's Rule with 2^n subdivisions. This program is not necessary for the
CASIO 7700 unless the user wants to compute and compare the values of the
midpoint and trapezoid rules with those obtained from Simpson's Rule.

CODE: INTEGRAL

Code (CASIO 7000)	Code (CASIO 7700)	Comments
"INTEGRAL"	INTEGRAL	Program title.
"A"? $\rightarrow$ A	"A"? $\rightarrow$ A	Input lower limit.
"B"? $\rightarrow$ B	"B"? $\rightarrow$ B	Input upper limit.
Lbl 1	Lbl 1	First label to Goto.
"N"? $\rightarrow$ N	"N"? $\rightarrow$ N	Input number of subintervals.
.5(B − A) ÷ N $\rightarrow$ H	.5(B − A) ÷ N $\rightarrow$ H	Determine half length of subintervals.
$\emptyset \rightarrow$ M	$\emptyset \rightarrow$ M	Initialize midpoint approximation M.
$\emptyset \rightarrow$ K	$\emptyset \rightarrow$ K	Initialize counter for loop.
Lbl 2	Lbl 2	Second label to Goto.
A+(2K + 1)H$\rightarrow$X	A+(2K + 1)H$\rightarrow$X	Determine midpoint of Kth subinterval.
Prog $\emptyset$:2HAns + M $\rightarrow$ M	2Hf$_1$ + M $\rightarrow$ M	Determine $f(X)$. Increment midpoint area M.
Isz K	Isz K	Increase K by 1.
K < N $\Rightarrow$ Goto 2	K < N $\Rightarrow$ Goto 2	If K<N, repeat from Lbl 2.
$\emptyset \rightarrow$ T	$\emptyset \rightarrow$ T	Initialize trapezoid approximation T.
$\emptyset \rightarrow$ K	$\emptyset \rightarrow$ K	Initialize counter for loop.
Lbl 3	Lbl 3	Third label to Goto.

(continued)

Code (CASIO 7000)	Code (CASIO 7700)	Comments
A + 2KH → X	A + 2KH → X	Determine left endpoint for Kth subinterval.
Prog ∅:HAns + T → T	Hf$_1$ + T → T	Determine $f(X)$. Add H$f(X)$ to T.
X + 2H → X	X + 2H → X	Determine right endpoint for Kth subinterval.
Prog ∅:HAns + T → T	Hf$_1$ + T → T	Determine $f(X)$. Add H$f(X)$ to T.
Isz K	Isz K	Increase K by 1.
K < N ⇨ Goto 3	K < N ⇨ Goto 3	If K<N, repeat from Lbl 3.
"M":M◢	"M":M◢	Display midpoint approximation M.
"T":T◢	"T":T◢	Display trapezoid approximation T.
"S":(2M + T)/3◢	"S":(2M + T)/3◢	Calculate and display Simpson's approximation.
Goto 1	Goto 1	Return to Lbl 1 and repeat.

OPERATION

Preparation

Enter the formula for the function to be numerically integrated into program ∅ (CASIO 7000) or f$_1$ (CASIO 7700). For example, to determine approximations for definite integrals of the function $f(x) = \sin(x^2)$, program ∅ or function f$_1$ must be entered as $\sin(X^2)$.

Input

When you run the program, input the following:

Prompt	Input
A?	Lower limit for the definite integral
B?	Upper limit for the definite integral
N?	Number of subintervals for the interval

Output

The program outputs various approximations for the definite integral in the following order:

M: The midpoint approximation using a Riemann sum with N subintervals

T: The trapezoid approximation using N subintervals

S: Simpson's approximation using 2N subintervals

Once M and T are determined, S is calculated by the formula $S = \dfrac{2M + T}{3}$.

The program does not terminate. Press the $\boxed{AC}$ key or press $\boxed{MODE}$ 1 to terminate INTEGRAL.

Sample Run

To use this program to approximate $\int_0^1 \sin (x^2)\, dx$ using N = 10 subintervals, enter program $\emptyset$ (CASIO 7000) or function f_1 (CASIO 7700) as sin (X^2). Run program INTEGRAL. Input A = 0, B = 1, and N = 10. You should get the following results:

M:	Midpoint approximation (10 subintervals):	0.3098162946
T:	Trapezoid approximation (10 subintervals):	0.3111708112
S:	Simpson's approximation (20 subintervals):	0.3102678001

A.12 Inverse

The program INVERSE sketches the graph of a function $y = f(x)$ on the viewing rectangle [1.5Ymin, 1.5Ymin] by [Ymin, Ymax] where Ymin and Ymax are chosen by the user. INVERSE then plots the graph of the inverse relation for a function. The viewing rectangle set by INVERSE uses the same scale for x and y. The program uses 98 (CASIO 7000) or 94 (CASIO 7700) bytes of memory.

CODE: INVERSE

Code (CASIO 7000)	Code (CASIO 7700)	Comments
"INVERSE"	INVERSE	Program title.
"A"? $\rightarrow$ A	"A"? $\rightarrow$ A	Input left endpoint of interval, store in A.
"B"? $\rightarrow$ B	"B"? $\rightarrow$ B	Input right endpoint of interval, store in B.
(B – A) ÷ 64 $\rightarrow$ Q	(B – A) ÷ 64 $\rightarrow$ Q	Determine step size. Store value in Q.
Int(10Q) $\rightarrow$ S	Int(10Q) $\rightarrow$ S	Determine tick marks, store value in S.
Range 1.5A, 1.5B, S, A, B, S	Range 1.5A, 1.5B, S, A, B, S	Set the Range to graph f and f^{-1}.

(continued)

Code (CASIO 7000)	Code (CASIO 7700)	Comments
Prog 1⏎	Graph Y = f_1⏎	Sketch the graph of function in Prog 1 or f_1.
Prog 2	Prog 2	Call program Cls-TXT to clear text screen.
Lbl 1	Lbl 1	First label for Goto.
A → X	A → X	Store the value of A in X.
Prog Ø:Plot Ans,A	Plot f_1,A	Find $f(A)$ and plot $(f(A), A)$.
A + Q → A	A + Q → A	Store A + Q in A.
A ≤ B ⇨ Goto 1	A ≤ B ⇨ Goto 1	If A ≤ B, repeat from Lbl 1. If not, end.
Plot	Plot	Keep graphic screen visible.

OPERATION

Preparation

Program INVERSE superimposes the graph of the inverse relation on the graph of the function. Choose Range settings for y to provide an appropriate viewing rectangle for the graph of the function and its inverse relation. Be sure program Cls-TXT is entered into program 2. (CASIO fx-7000GA calculator does not require Cls-TXT. Omit line **Prog 2** from program code.)

CASIO 7000

Enter the formula for the function $y = f(x)$ into program Ø. Enter the graph statement for f into program 1. For example, to plot the graph of the inverse relation for the function $f(x) = x^2$, enter program Ø as X^2 and program 1 as Graph Y = X^2.

CASIO 7700

Enter the formula for the function $y = f(x)$ into function f_1 in the function menu. For example, to plot the graph of the inverse relation for the function $f(x) = x^2$, enter f_1 as X^2.

Input

When you run the program, input the following:

Prompt   Input

A?   The left endpoint of the interval that gives the range of the function and the domain of its inverse

B?   The right endpoint of the interval that gives the range of the function and the domain of its inverse

Choose A to be less than B. The program terminates without plotting the graph of f^{-1} if B is chosen to be greater than A.

Output

The program will sketch the graph of the function $y = f(x)$ stored in programs $\emptyset$ and 1 (CASIO 7000) or f_1 (CASIO 7700). INVERSE then plots the graph of the inverse relation $(f(x), x)$ for this function on a viewing rectangle whose x and y scales are the same. The graphs of the function and its inverse relation are displayed on the screen when the program terminates. Press $\boxed{\text{EXE}}$ to start the program again.

Sample Run

Preparation

To sketch a graph of the function $f(x) = x^2$ and its inverse relation on a viewing rectangle [–4.5, 4.5] by [–3, 3], complete the following.

CASIO 7000

Enter program $\emptyset$ as X^2. Store the graph statement Graph Y = X^2 in program 1.

CASIO 7700

Enter X^2 into function f_1.

Run

Run program INVERSE and input A = –3, B = 3. This results in the display of the graphs of the parabolas $y = x^2$ and $x = y^2$ on the screen.

A.13 Newton's Method

NEWTON'S METHOD-GRAPHICAL

Program NEWTON.G uses a modification of Newton's Method to approximate a zero of a function f. If c is the first approximation of a zero of f, Newton's Method obtains the next approximation as the x-intercept of the line tangent to the graph of $y = f(x)$ at the point $(c, f(c))$. This next approximation b is obtained by the formula

$$b = c - \frac{f(c)}{f'(c)}$$

The modification of Newton's Method is that a numerical approximation is used for the value of the derivative $f'(c)$. The approximation is

$$f'(c) \approx \frac{f(c + h) - f(c - h)}{2h}$$

where $h = 10^{-6}$. This approximation for the derivative is obtained using the subroutine NDERIV (page 189). NEWTON.G uses 116 (CASIO 7000) or 113 (CASIO 7700) bytes of memory. NDERIV requires 38 (CASIO 7000) or 32 (CASIO 7700) bytes of memory.

CODE: NEWTON.G

Code (CASIO 7000)	Code (CASIO 7700)	Comments
"NEWTON.G"	NEWTON.G	Program title.
Cls	Cls	Clear the graphics screen.
Prog 1◢	Graph Y = f1◢	Sketch the graph of $y = f(x)$.
"C" ? → C	"C" ? → C	Input C. Store in C.
E–6 → H	E–6 → H	Initialize H.
Lbl 1	Lbl 1	First label to Goto.
C → X	C → X	Store C in X.
"F(C)"	"F(C)"	Display F(C).
Prog Ø:Ans → F◢	f_1 → F◢	Store $f(C)$ in F.
Prog 3	Prog 3	Call program NDERIV to determine $f'(C)$.
Plot C, Ø:Plot C,F:Line	Plot C, Ø:Plot C,F:Line	Draw vertical line from x-intercept to graph of f.
Graph Y = M(X–C)+F◢	Graph Y = M(X–C)+F◢	Graph tangent at $(C, f(C))$.
C – F÷M → R	C – F÷M → R	Calculate next approximation, store in R.
Abs(C–R) < E–3 ⇨ Goto 2	Abs(C–R) < E–3 ⇨ Goto 2	Compare successive values. If close, end program.
"C":R → C◢	"C":R → C◢	Display approximation C.
Goto 1	Goto 1	Repeat from Lbl 1 with next approximation.
Lbl 2	Lbl 2	End of program label.
"R":R	"R":R	Display approximate root.

OPERATION

Preparation

To use this program, set the Range to provide an appropriate viewing rectangle that displays a root of the function. Enter the subroutine code for program NDERIV into program 3. Note: If NDERIV is entered into a different numbered program location, be sure to enter that program number into the code for NEWTON in place of the code **Prog 3**.

CASIO 7000

Store the formula for the function in program $\emptyset$. Store the graph statement for the function in program 1. For example, if the function is $f(x) = x^2 - 10$, then program $\emptyset$ must be entered as $X^2 - 10$. Enter program 1 as Graph Y = $X^2 - 10$.

CASIO 7700

Store the formula for the function in function f_1. For example, if the function is $f(x) = x^2 - 10$, then function f_1 must be entered as $X^2 - 10$.

Input

When the program is run, the graph of the function stored in programs $\emptyset$ and 1 (CASIO 7000) or f_1 (CASIO 7700) is displayed. Press $\boxed{\text{EXE}}$. Input the initial approximation for the zero of the function at the prompt **C?**. The initial approximation is stored in C.

Output

The program first sketches the graph of the function stored program 1 or function f_1. It outputs the values of C and $f(C)$. NEWTON.G then sketches the graph of a vertical line from the x-intercept $(C, 0)$ of the previous tangent to the graph of f. The tangent line is drawn to intersect f at $(C, f(C))$. After each output line or graph, press $\boxed{\text{EXE}}$ to continue. The program terminates when the absolute value of the difference between two successive approximations is less than 10^{-3}.

Sample Run

Preparation

To use the program NEWTON.G to find a positive zero of $f(x) = x^2 - 10$, set the Range to graph f on a viewing rectangle [0, 6] by [−10, 21]. Enter the subroutine code for program NDERIV into program 3. Note: If NDERIV is entered into a different numbered program location, be sure to enter that program number into the code for TANGENT in place of the code **Prog 3**.

CASIO 7000

Enter $X^2 - 10$ into program $\emptyset$ and Graph Y = $X^2 - 10$ into program 1.

CASIO 7700

Enter $X^2 - 10$ into function f_1.

Run

Run NEWTON.G using C = 1 as the initial approximation. (This allows a few separate tangent lines to be shown.) Successive values of C and $f(C)$ are displayed, followed by the graphs of a vertical line through the x-intercept of the tangent line to the graph of f and the tangent line through $(C, f(C))$. When the program terminates, it shows an approximate root of R = 3.162277665.

Note: It is possible to convert this program to the program NEWTON, which does not show the graphical interpretation of Newton's Method.

NEWTON'S METHOD

Program NEWTON is a modification of program NEWTON.G. The lines of code that provide a graphical illustration of Newton's Method are deleted. NEWTON provides a numerical approximation for a zero of a function $y = f(x)$. If a is the first approximation of a zero of f, NEWTON obtains the next approximation as the x-intercept of the line tangent to the graph of $y = f(x)$ at the point $(a, f(a))$. This program uses 65 (CASIO 7700) or 62 (CASIO 7700) bytes of memory.

CODE: NEWTON

The code for this program can be obtained by deleting several lines and changing four lines of the code for the program NEWTON.G. The final code is as follows:

Code (CASIO 7000)	Code (CASIO 7700)	Comments
"NEWTON"	NEWTON	Program title.
E – 6 → H	E – 6 → H	Initialize H.
Lbl 1	Lbl 1	First label to Goto.
Prog $\emptyset$:Ans → F◢	f_1 → F◢	Store f(C) in F.
Prog 3	Prog 3	Call program NDERIV to determine f'(C).
X – F ÷ M → R◢	X – F ÷ M → R◢	Calculate and display next approximation; store in R.
Abs(X – R) < E–6 ⯈ Goto 2	Abs(X – R) < E–6 ⯈ Goto 2	Compare successive values. If close, end program.
R → X	R → X	Store R in X.
Goto 1	Goto 1	Repeat from Lbl 1 with next approximation.
Lbl 2	Lbl 2	End of program label.
"R":R	"R":R	Display approximate root.

OPERATION

Preparation

To use this program, sketch an appropriate graph of $y = f(x)$ on the calculator before running the program. Enter the subroutine code for program NDERIV into program 3. Note: If NDERIV is entered into a different numbered program location, be sure

to enter that program number into the code for NEWTON in place of the code
Prog 3.

CASIO 7000

Store the formula for the function in program $\emptyset$. For example, if the function is
$f(x) = x^2 - 10$, then program $\emptyset$ must be entered as $X^2 - 10$.

CASIO 7700

Store the formula for the function in function f_1. For example, if the function is
$f(x) = x^2 - 10$, then function f_1 must be entered as $X^2 - 10$.

Input

Before running the program, input the initial approximation for the zero of the
function from the graph of the function by using the trace option of the calculator.
Sketch a graph of the function that shows the zero of the function as an x-intercept
of the graph. Then use the trace function to trace along the graph until the blinking
cursor is close to the desired x-intercept. The x-coordinate of the point obtained
from the trace will serve as the initial approximation for the zero of the function.
Run NEWTON.

Output

The program displays successive approximations of the root of the equation
determined by Newton's Method. The program runs until the absolute value of the
difference between two successive approximations is less than 10^{-6}. The program
outputs the approximation for the zero as R.

To run the program again to approximate another zero of the same function, use the
Trace key as before to obtain an initial approximation for the other zero.

Sample Run

To use the program NEWTON to find a positive zero of $f(x) = x^2 - 10$, first enter
$X^2 - 10$ into program $\emptyset$ (CASIO 7000) or function f_1 (CASIO 7700). Next, graph
$y = x^2 - 10$ on [–5, 5] by [–10, 5]. Then use the trace option to trace close to the
x-intercept between 3 and 4. Finally, run the program NEWTON. The output
should be R = 3.16227866.

Alternate Use

Program NEWTON can be used with an initial approximation that is input
numerically rather than from the graph of $y = f(x)$. Store the value of the initial
approximation for the root in X. Run NEWTON as before to find a numerical
approximation for the root.

ALTERNATE VERSION

Program NEWTON.A is an adaptation of program NEWTON. NEWTON.A uses
Newton's Method to approximate a zero of a function f when the value of the initial
approximation is entered numerically rather than from a graph of the function.

NEWTON.A also uses the formula for the first derivative of f rather than a numerical approximation for the value of $f'(a)$, allowing the approximation for the root to be more accurate. This program uses 59 (CASIO 7000) or 55 (CASIO 7700) bytes of memory.

CODE: NEWTON.A

Code (CASIO 7000)	Code (CASIO 7700)	Comments
"NEWTON.A"	NEWTON.A	Program title.
"A" $\rightarrow$ A	"A" $\rightarrow$ A	Input A. Store in A.
Lbl 1	Lbl 1	First label to Goto.
Prog $\emptyset$		Store $f(C)$ in F.
X − F ÷ M $\rightarrow$ R	X − f_1 ÷ f_2 $\rightarrow$ R	Calculate and display next approximation; store in R.
Abs(X − R) < E − 10 ⇨ Goto 2	Abs(X − R) < E−10 ⇨ Goto 2	Compare successive values. If close, end program.
R $\rightarrow$ X	R $\rightarrow$ X	Store R in X.
Goto 1	Goto 1	Repeat from Lbl 1 with next approximation.
Lbl 2	Lbl 2	End of program label.
"R":R	"R":R	Display approximate root.

OPERATION

Preparation

CASIO 7000

To use this program, store the formula for the function and the formula for the first derivative in program zero as follows:

$$f(X) \rightarrow F : f'(X) \rightarrow M$$

For example, if the function is $f(x) = x^2 - 10$, then enter program $\emptyset$ as

$$X^2 - 10 \rightarrow F : 2X \rightarrow M$$

CASIO 7700

To use this program, store the formula for the function in function f_1 and the formula for the derivative of the function in function f_2. For example, if the function is $f(x) = x^2 - 10$, enter function f_1 and function f_2 as $f_1 = X^2 - 10$ and $f_2 = 2X$, respectively.

Input

When you run the program, input the initial approximation for the zero of the function at the prompt **A?**. The initial approximation is stored in X.

Output

The program displays the root of the equation determined by Newton's Method. NEWTON.A runs until the absolute value of the difference between two successive approximations is less than 10^{-10}. The program outputs the approximation for the zero as R.

Sample Run

Preparation
Use the program NEWTON.A to find a positive zero of $f(x) = x^2 - 10$ as follows:

CASIO 7000

Enter program $\emptyset$ as follows:

$$X^2 - 10 \rightarrow F : 2X \rightarrow M$$

CASIO 7700

Enter $X^2 - 10$ into function f_1 and $2X$ into function f_2.

Run
Run this program using A = 3 as the initial approximation. The approximate root is R = 3.16227766.

NEWTON'S METHOD (Using the $\boxed{\text{Ans}}$ key)

The value of the last computation can be obtained by pressing the $\boxed{\text{Ans}}$ key. For example, calculate $5 + 2$ on your calculator. Then press

$$\boxed{\text{Ans}} \; \boxed{+} \; 8 \; \boxed{\text{EXE}}$$

The number 15 should appear $(7 + 8 = 15)$.

This makes the $\boxed{\text{Ans}}$ key very useful in any recursive computation such as Newton's Method. If x_n is an approximation for the zero of a function, then the next approximation is obtained from x_n according to the formula

$$x_{n+1} = x_n - \frac{f(x_n)}{f'(x_n)}$$

To perform Newton's Method directly on the CASIO, enter the initial approximation to the zero of the function. Press $\boxed{\text{EXE}}$ followed by the code

$$\text{Ans} - f(\text{Ans}) \div f'(\text{Ans})$$

Pressing the $\boxed{\text{EXE}}$ key repeatedly will produce successive approximations to the zero of the function.

It is necessary to enter the formula for the function, $f(\text{Ans})$, and the formula for the derivative of the function $f'(\text{Ans})$. For example, to approximate the positive zero of the function $f(x) = x^2 - 10$ using 3 as the initial approximation, press

$$3 \ \boxed{\text{EXE}}$$

and then enter the code

$$\text{Ans} - (\text{Ans}^2 - 10) \div (2\text{Ans}).$$

Press the $\boxed{\text{EXE}}$ key several times. When two successive approximations are identical on the calculator, you are finished. In this case, 3.16227766 should be obtained as the approximate zero of this function.

A.14 Numerical Derivative

Program NDERIV numerically approximates the value of the derivative of a function $y = f(x)$ at a point $x = a$. NDERIV uses the following expression to approximate the value of the derivative at $x = a$:

$$f'(a) = \frac{f(a + h) - f(a - h)}{2h}$$

where $h = 10^{-6}$. NDERIV requires 65 (CASIO 7000) or 59 (CASIO 7700) bytes of memory.

CODE: NDERIV

Code (CASIO 7000)	Code (CASIO 7700)	Comments
"A"? $\rightarrow$ A	"A"? $\rightarrow$ A	Input A, store in A.
E $-$ 6 $\rightarrow$ H	E $-$ 6 $\rightarrow$ H	Store H as 10^{-6}.
A $+$ H $\rightarrow$ X	A $+$ H $\rightarrow$ X	Store value of A $+$ H in X.

(continued)

Code (CASIO 7000)	Code (CASIO 7700)	Comments
Prog Ø: Ans → F	f_1 → F	Determine $f(A + H)$. Store value in F.
A – H → X	A – H → X	Store value of A – H in X.
Prog Ø: Ans → G	f_1 → G	Determine $f(A – H)$. Store value in G.
"NDERIV":E6(F – G) ÷ 2	"NDERIV":E6(F–G) ÷ 2	Display NDERIV value.

OPERATION

Preparation

Store the formula for the function for which the value of the numerical derivative is desired in program Ø (CASIO 7000) or function f_1 (CASIO 7700). For example, if the value of the derivative at $x = 2$ is desired for the function $f(x) = x^2$, store X^2 in program Ø or f_1.

Input

Enter the value of $x = a$ at the **A?** prompt. No other input is required.

Output

The numerical approximation for the derivative NDERIV is displayed. The numerical approximation is the estimate of the derivative at $x = a$ given by the expression

$$f\,'(a) = \frac{f(a + h) - f(a - h)}{2h}$$

where $h = 10^{-6}$ and $y = f(x)$ is the function stored in program Ø (CASIO 7000) or function f_1 (CASIO 7700).

Sample Run

To determine the numerical derivative of the function $y = \sqrt{1 + x^2}$ at $x = 2$, enter the formula $\sqrt{(1 + x^2)}$ into program Ø (CASIO 7000) or function f_1 (CASIO 7700). Run NDERIV, entering the value of A as 2 at the **A?** prompt. The numerical approximation for NDERIV is given as 0.894427.

Subroutine Adaptation

Program NDERIV is useful as a subroutine for programs that use the value of the numerical derivative, such as NEWTON and TANGENT. When NDERIV is used as a subroutine of another program, the calling programs will supply the values of A and H. NDERIV outputs the value of the numerical derivative, storing this value in

M. When this program is called from NEWTON and TANGENT, it is referred to as program 3. Store subroutine NDERIV in program location 3, or change the number in the program code **Prog 3** to correspond with the program location of NDERIV.

It is possible to use the subroutine code directly to find a single value for the numerical derivative for a function. To do this, first store the values of X and H into these memory locations. For example, to estimate the slope of the tangent to the graph of $f(x) = x^3 - 4$ at $x = 2$, store the function in program $\emptyset$ (CASIO 7000) or function f_1 (CASIO 7700), store 2 in X and E – 6 in H. Run NDERIV. The value of the numerical derivative is given as 12.

Subroutine code: NDERIV

Code (CASIO 7000)	Code (CASIO 7700)	Comments
$X \rightarrow A$	$X \rightarrow A$	Input X, store in A.
$A - H \rightarrow X$	$A - H \rightarrow X$	Store value of A – H in X.
Prog $\emptyset$: Ans $\rightarrow$ G	$f_1 \rightarrow G$	Determine $f(A - H)$. Store value in G.
$A + H \rightarrow X$	$A + H \rightarrow X$	Store value of A + H in X.
Prog $\emptyset$		Determine $f(A + H)$.
(Ans – G) ÷ (2H) $\rightarrow$ M	(Ans – G) ÷ (2H) $\rightarrow$ M	Determine value of NDERIV. Store in M.

A.15 Parametric Grapher

Program PARAM graphs curves defined parametrically as $(x(t), y(t))$ where x and y are functions of t. The user inputs the t-interval over which the graph is to be sketched. Program PARAM is included for use with CASIO 7000 series calculators. CASIO 7700 calculators have a built-in parametric grapher, so it has no need for this program. PARAM requires 69 bytes of memory.

CODE: PARAM

Code (CASIO 7000)	Comments
"PARAM"	Title of program.
"TMIN"? $\rightarrow$ T	Input Tmin and store in T.
"TMAX"? $\rightarrow$ E	Input Tmax and store in E.
"N"? $\rightarrow$ N	Input number of points to be sketched.

(*continued*)

Code (CASIO 7000)	Comments
$(E - T) \div N \to S$	Determine step size for T.
Prog 2	Call program Cls-TEXT as subroutine.
Lbl 1	First Lbl to Goto.
Prog $\emptyset$	Call program $\emptyset$ for function evaluation.
Plot X, Y	Plot the points $(x(t), y(t))$.
Line	Connect graph (rather than plotting by points).
$T + S \to T$	Increment T.
$T \leq E \Rightarrow$ Goto 1	Test if T is still within t-interval.
Plot	Display graph when T is larger than E.

OPERATION

Preparation

Store the functions $x = x(t)$ and $y = y(t)$ in program $\emptyset$ as follows:

$$x(t) \to X: y(t) \to Y.$$

Set the Range appropriately to graph the curve. Store program Cls-TEXT in program 2 (this is not necessary for the CASIO fx-7000GA).

Input

Run program PARAM. At the appropriate prompts, input Tmin, Tmax, and the number of points N to be plotted.

Output

The parametric curve is plotted for the t-interval chosen.

Sample Run

To graph the parametric curve $(\cos t, 2 \sin t)$, set the viewing rectangle to graph the curve on $[-4.7, 4.7]$ by $[-3.1, 3.1]$. Store program $\emptyset$ as follows:

$$\cos T \to X: 2\sin T \to Y$$

Run PARAM. Choose Tmin = 0, Tmax = 2π, and N = 50. Press $\boxed{\text{EXE}}$ to sketch the curve. While the curve is being sketched, the screen appears to flicker. By watching carefully, you should be able to see the graph as it is being sketched.

Adaptation

It is possible to graph polar functions using PARAM. To graph $r = f(\theta)$, store program $\emptyset$ as follows: $f(T) \to R: R \cos T \to X: R \sin T \to Y$. Then run PARAM as described above. For example, to graph $r = \sin \theta$, store program $\emptyset$ as

$$\sin T \to R: R \cos T \to X: R \sin T \to Y$$

A.16 Polar Families

Program POLAR allows the simultaneous display of several graphs of the family of functions $r = A f(B\theta - \frac{C\pi}{4}) + D$ where A, B, C, and D are chosen by the user.

With POLAR, the T-interval is set from Tmin to Eπ where Tmin and E are chosen by the user. POLAR requires 116 (CASIO 7000) or 38 (CASIO 7700) bytes of memory. Since the code and operation of POLAR for the CASIO 7000 and the CASIO 7700 are quite different, these programs are listed separately.

CODE: POLAR (CASIO 7000)

Code (CASIO 7000)	Comments
"POLAR"	Title of program.
Cls	Clear graphics screen.
"TMIN"? $\rightarrow$ F	Input Tmin and store in T.
"E"? $\rightarrow$ E	Input E and store in E.
"N"? $\rightarrow$ N	Input number of points to be sketched.
(Eπ – F) ÷ N $\rightarrow$ S	Determine step size for T.
Lbl 1	First Lbl to Goto.
F $\rightarrow$ T	Store the initial value of T in T.
"A"? $\rightarrow$ A	Input A.
"B"? $\rightarrow$ B	Input B.
"C"? $\rightarrow$ C	Input C.
"D"? $\rightarrow$ D	Input D.
Prog 2	Call program Cls-TEXT as a subroutine to clear the text screen. Delete this line for fx-7000GA.
Lbl 2	Second Lbl to Goto.
Prog ∅	Call program ∅ for function evaluation.
Plot R cos T, R sin T	Plot the point (R cos T, R sin T).
T + S $\rightarrow$ T	Increment T.
T $\leq$ Eπ ⇨ Goto 2	If T is still within t-interval, repeat from Lbl 2.
Plot◢	Display graph and pause.
Goto 1	Repeat the program from Lbl 1.

OPERATION

Preparation

Store the function $r = A f\left(B\theta - \dfrac{C\pi}{4}\right) + D$ in program $\emptyset$ as follows:

$$A f(BT - \frac{C\pi}{4}) + D \rightarrow R$$

Set the Range appropriately to graph the family of polar functions. Use program SQ.SCREEN (page 205) or use a multiple of the default viewing window to make certain that x and y intervals have the same scale. It is important that circles appear as circles and ellipses as ellipses.

Input

At the appropriate prompts, input Tmin, E where Eπ is Tmax, and the number N of points to be plotted. Also input the values of A, B, C, and D to graph the appropriate family member of $r = A\ f(BT - \dfrac{C\pi}{4}) + D$.

Output

The graph of the polar function determined by the choices of A, B, C, and D is sketched. Pressing $\boxed{\text{EXE}}$ upon completion of the graph runs POLAR again. This allows multiple members of a single polar family to be displayed on the same viewing rectangle.

Sample Run

To graph several members of the family $r = A\ \sin(B\theta - \dfrac{C\pi}{4}) + D$, store the expression A sin(BT – Cπ/4) + D $\rightarrow$ R in program $\emptyset$. Set the Range to [–4.7, 4.7] by [–3.1, 3.1]. Run POLAR. At the appropriate prompt, enter Tmin as 0, E as 2, and N as 50. Graph r = sin T by letting A = 1, B = 1, C = 0, and D = 0. Press $\boxed{\text{EXE}}$ to run POLAR again. Graph r = 2 sin T by letting A = 2, B = 1, C = 0, and D = 0. Experiment with other values of A, B, C, and D.

CODE: POLAR (CASIO 7700)

Before storing the program, set the mode of the calculator to polar by pressing $\boxed{\text{MODE}}$ $\boxed{\text{SHIFT}}$ to display the graph type mode screen. Press $\boxed{-}$ to change the graphing mode to polar.

Code (CASIO 7700)	Comments
'FAMILY	Title of program.
"A"? $\rightarrow$ A	Input A.
"B"? $\rightarrow$ B	Input B.
"C"? $\rightarrow$ C	Input C.
"D"? $\rightarrow$ D	Input D.
Graph r = f$_1$	Draw graph of $r = A \cdot f(BT - C\pi \div 4) + D$.

OPERATION

Preparation

With the graphing mode set to polar, insert $r = f(\theta)$ into function f$_1$ in the function memory as follows: $A f(B\theta - C\pi \div 4) + D$. Set the Range appropriately to graph the family of functions. If it is desirable to use the t-interval [0, 2π], press $\boxed{\text{F1}}$ to INITialize the Range. This also changes the x and y intervals. Otherwise, store the Range values as usual (note that it is possible to store multiples of π in Tmax without having to first determine a numerical approximation for π).

Input

At the appropriate prompts, input the values of A, B, C, and D to graph the appropriate family member of $r = A f(BT - \dfrac{C\pi}{4}) + D$.

Output

The graph of the polar function determined by the choices of A, B, C, and D is sketched. Pressing $\boxed{\text{EXE}}$ upon completion of the graph runs POLAR again. This allows multiple members of a single polar family to be displayed on the same viewing rectangle.

Sample Run

To graph several members of the family $r = A \sin(B\theta - \dfrac{C\pi}{4}) + D$, change the calculator mode to **Polar**. Store the expression $A \sin(B\theta - C\pi \div 4) + D$ in f$_1$ in the function menu. Initialize the Range. Run POLAR. At the appropriate prompt, graph $r = \sin T$ by letting A = 1, B = 1, C = 0, and D = 0. Press $\boxed{\text{EXE}}$ to run POLAR again. Graph $r = 2 \sin T$ by letting A = 2, B = 1, C = 0, and D = 0. Experiment with other values of A, B, C, and D.

A.17 Riemann

The program RIEMANN determines the left endpoint and right endpoint Riemann sum approximations for the definite integral of a function f over an interval $[a, b]$. If the function f is positive on the interval $[a, b]$, then these Riemann sums will approximate the area under the graph of the function over the interval $[a, b]$. This program requires 197 (CASIO 7000) or 195 (CASIO 7700) bytes of memory.

CODE: RIEMANN

Code (CASIO 7000)	Code (CASIO 7700)	Comments
"RIEMANN"	RIEMANN	Program title.
"A"? → A	"A"? → A	Input value of lower limit of integration. Store in A.
"B"? → B	"B"? → B	Input value of upper limit of integration. Store in B.
Lbl 1	Lbl 1	First label to Goto.
"N"? → N	"N"? → N	Input number of sub-intervals. Store in N.
(B – A) ÷ N → H	(B – A) ÷ N → H	Determine length of subintervals.
"L OR R"? → W	"L OR R"? → W	Input a letter L for left, R for right. Store in W.
∅ → S	∅ → S	Initialize sum S to zero.
Cls	Cls	Clear graphics screen.
Prog 1◢	Graph Y = f_1◢	Draw graph in Prog 1 or f_1.
Prog 2	Prog 2	Call program Cls-TEXT to clear text screen.
W ≠ L ⇨ Goto2	W ≠ L ⇨ Goto2	Test for right endpoints. Goto Lbl 2 if W≠L.
N – 1 → T	N – 1 → T	Store value of N – 1 in T.
∅ → V	∅ → V	Initialize V to zero.
H → Z	H → Z	Store the value of H in Z.
Lbl 3	Lbl 3	Third Lbl to Goto.
A + VH → X	A + VH → X	Determine value of partition point.
Prog ∅:Ans → Q	f_1 → Q	Determine height of rectangle at partition point.

(continued)

Code (CASIO 7000)	Code (CASIO 7700)	Comments
S + QH → S	S + QH → S	Increment sum by area of rectangle.
X → P	X → P	Store the value of X in P.
Plot P, Ø	Plot P, Ø	Plot point P, Ø.
Plot P, Q:Line	Plot P, Q:Line	Draw part of approximating rectangle.
Plot P + Z, Q:Line	Plot P + Z,Q:Line	Draw part of approximating rectangle.
Plot P + Z, Ø:Line	Plot P + Z, Ø:Line	Draw part of approximating rectangle.
V + 1 → V	V + 1 → V	Increase V by 1. Store in V.
V≤T ⇨ Goto 3	V≤T ⇨ Goto 3	Return control of program to Lbl 3 if V≤T.
Plot◢	Plot◢	Pause with graphics screen visible.
"SUM":S◢	"SUM":S◢	Display the SUM and its value.
Goto 1	Goto 1	Return control of the program to Lbl 1.
Lbl 2	Lbl 2	Second place to Goto.
W ≠ R ⇨ Goto 1	W ≠ R ⇨ Goto 1	Test for left endpoints. Goto Lbl 2 if W ≠ R.
N → T	N → T	Store the value of N in T.
1 → V	1 → V	Store one in V.
–H → Z	–H → Z	Store the value of –H in Z.
Goto 3	Goto 3	Return control of the program to Lbl 3.

OPERATION

Preparation

Store program Cls-TEXT in program 2. (CASIO fx-7000GA calculator does not require Cls-TXT; omit line **Prog 2** from RIEMANN code.)

<u>CASIO 7000</u> Store the formula for the function to be used in program Ø. Store the graph statement for the function in program 1. For example, if the function is $f(x) = x^2$, then enter program Ø as X^2. Enter program 1 as Graph Y = X^2. Sketch

the graph by running program 1 to determine appropriate Range settings to display a suitable graph.

<u>CASIO 7700</u> Store the formula for the function to be used in function f_1. For example, if the function is $f(x) = x^2$, then enter function f_1 as X^2. Sketch the graph of the function in f_1 to determine appropriate Range settings to display a suitable graph.

Input

When you run the program, input the following:

Prompt	Input
A?	Lower limit of integration for the definite integral
B?	Upper limit of integration for the definite integral
N?	Number of subdivisions for the interval
L OR R?	L for rectangles whose height is determined by the function value at the subintervals' left endpoints, or R for rectangles whose height is determined by the function value at the subintervals' right endpoints

CAUTION: When B is chosen to be less than A, choosing L generates rectangles whose height is determined by the function value at the right endpoints and R gives left-endpoint information. If N is chosen to be negative, interval widths are negative, resulting in a change of sign for the value of the sum.

Output

The program first sketches the graph of the function using the Range settings previously set by the user. The interval [A, B] is divided into N equal subintervals. Choosing L results in a sketch of the approximating rectangles for the left endpoint approximation. When the $\boxed{\text{EXE}}$ key is pressed, the value of the left endpoint approximation is displayed. Choosing R results in a sketch of the right endpoint rectangles followed by the right endpoint approximation. Pressing $\boxed{\text{EXE}}$ allows the program to be repeated using the same function and the same interval. Input a new value for N and choose L (left) or R (right) once again.

RIEMANN does not terminate. Press $\boxed{\text{AC}}$ to terminate the program when a graph is visible or immediately after the SUM is displayed. Terminate by pressing $\boxed{\text{MODE}}$ 1 otherwise.

Sample Run

Preparation

To use this program to approximate the area of the graph of $f(x) = x^2$ between $x = 0$ and $x = 2$ with Riemann sum approximations, set the Range settings to

Xmin = −.5	Xmax = 2.5	Scale = 1
Ymin = −1	Ymax = 6	Scale = 1

<u>CASIO 7000</u> Store X^2 in program ∅. Store Graph $Y = X^2$ in program 1.

<u>CASIO 7700</u> Store X^2 in function f_1.

Run

Run program RIEMANN. To obtain the left-endpoint approximation for four subintervals, choose A = 0, B = 2, N = 4, and L. After the rectangles are drawn for the left endpoint approximation, the program will display SUM = 1.75. Press $\boxed{\text{EXE}}$. Choose N = 4 and R to see the rectangles for the right endpoint approximation. Press $\boxed{\text{EXE}}$ once more to display R = 3.75. After pressing $\boxed{\text{EXE}}$ again, input N = 8. Run RIEMANN to display the graphs and corresponding left and right endpoint approximations. The values for left and right in this case are 2.1875 and 3.1875, respectively.

Notice that since $f(x) = x^2$ is increasing on the interval [0, 2], left endpoints L result in a lower approximating sum for this area, and right endpoints R give an upper approximating sum for this area.

A.18 Secant

Program SECANT draws secant lines for the graph of a function f through the points $(c, f(c))$ and $(c+h, f(c+h))$. It also calculates the slope of each secant line that is drawn. These secant lines can be used to approximate a tangent line to the graph of f at the point $(c, f(c))$. The slopes of the secant lines approximate the slope of the tangent line for small values of h. This program requires 102 (CASIO 7000) or 97 (CASIO 7700) bytes of memory.

CODE: SECANT

Code (CASIO 7000)	Code (CASIO 7700)	Comments
"SECANT"	SECANT	Title of program.
Cls	Cls	Clear graphics screen.
Prog 1⁂	Graph Y = f_1⁂	Graph $y = f(x)$.
"C"? → C	"C"? → C	Input C. Store value in C.
"H"? → H	"H"? → H	Input H. Store value in H.
C → X	C → X	Copy value of C in X.
Prog ∅:Ans → B	f_1 → B	Calculate $f(C)$. Store in B.
Lbl 1	Lbl 1	First label for Goto.

(continued)

Code (CASIO 7000)	Code (CASIO 7700)	Comments
C + H → X	C + H → X	Store C + H in X.
Prog Ø	f$_1$	Evaluate f(C + H).
"M"	"M"	Display M.
(Ans – B) ÷ H → M◢	(Ans – B) ÷ H → M◢	Calculate and display slope.
Graph Y = M(X – C)+B◢	Graph Y = M(X – C)+B◢	Graph secant line.
H ÷ 2 → H	H ÷ 2 → H	Replace H with H ÷ 2.
Abs H < .01 ⇨ Goto 2	Abs H < .01 ⇨ Goto 2	If |H|<.01 end program.
"H":H◢	"H":H◢	Display H and value of H.
Goto 1	Goto 1	Repeat from Lbl 1.
Lbl 2	Lbl 2	End of program.

OPERATION

Preparation

Select an appropriate viewing rectangle for the graph and set the Range.

<u>CASIO 7000</u> Store the formula for the function to be used in program Ø. Store the graph statement for the function in program 1. For example, to use the function $f(x) = \sin x$, enter the code sin X into program Ø. Enter the code into program 1 as Graph Y = sin X.

<u>CASIO 7700</u> Store the formula for the function to be used in function f$_1$. For example, to use the function $f(x) = \sin x$, function f$_1$ must be stored as sin X.

Input

The program draws secant lines for the graph of the function connecting the points $(C, f(C))$ and $(C + H, f(C + H))$. Press EXE after the graph of f has been displayed. Input the value of C and the initial value of H at the appropriate prompts.

Output

The program first draws the graph of the function stored in program 1 or in function f$_1$ on the selected viewing rectangle. Displayed next is the value of the slope (M) of the secant line connecting the points $(C, f(C))$ and $(C + H, f(C + H))$ for the values of C and H selected by the user. Pressing EXE once more, SECANT sketches the secant line on the graph of the function. The program replaces H with $\frac{H}{2}$, displays the value of H and repeats the process. Continue pressing $\boxed{\text{EXE}}$ to see the values of successive slopes M, interval widths H, and resulting secant lines to the graph of f through the points $(C, f(C))$ and $(C + H, f(C + H))$. The procedure is repeated until $|H| < 0.01$, at which time the program ends. With this program, the limit of the slope of the secant lines as the value of H gets closer and closer to zero can be investigated.

Sample Run

Preparation

To investigate the slope of secant lines for the function $f(x) = \sin x$ at the point $(\frac{\pi}{2}, 1)$, set the Range so that Xmin = 0, Xmax = 5, Xscl = 1, Ymin = –2, Ymax = 2, Yscl = 1.

<u>CASIO 7000</u> Store the formula for $f(x) = \sin x$ in program $\emptyset$ as sin X. Store the graphs statement for this function in program 1 as Graph Y = sin X.

<u>CASIO 7700</u> Store the formula for $f(x) = \sin x$ in function f_1 as sin X.

Run

Run program SECANT with $C = \frac{\pi}{2}$ and H = 2. Press the $\boxed{\text{EXE}}$ key until four secant lines are drawn. Make note of the values of H and M (the slope of the secant line) as the program progresses. The fourth secant line will have H = .25 and M = –0.1243503132. Continue pressing the $\boxed{\text{EXE}}$ key until the program terminates.

Adaptation

Program SECANT can be used to sketch the graph of a function $y = f(x)$ and a single tangent line to the graph of f at a point $(c, f(c))$. Enter the desired function as described for SECANT. Set the Range. Run SECANT and enter the value of c at the C? prompt. Enter H as some small value such as .01. The graph of f is sketched followed by a secant line between two points which are so close together that they coincide on the graphics screen, providing the appearance of the graph of the tangent to f at $(c, f(c))$. Press $\boxed{\text{AC}}$ to terminate the program.

A.19 Series

The program SERIES can be used to find a partial sum of a given series. If a_k is the kth term of the series, this program can be used to determine the sum $\sum_{k=1}^{N} a_k$.

This program uses 59 (CASIO 7000) or 55 (CASIO 7700) bytes of memory.

CODE: SERIES

Code (CASIO 7000)	Code (CASIO 7700)	Comments
"SERIES"	SERIES	Title of program.
"I"? → K	"I"? → K	Input initial index value. Store in K.
"N"? → N	"N"? → N	Input final index value. Store in N.
$\emptyset$ → S	$\emptyset$ → S	Initialize sum.
Lbl 1	Lbl 1	First label for Goto.
Prog $\emptyset$: Ans + S → S	f_1 + S → S	Calculate term of series and add term to sum S.
K + 1 → K	K + 1 → K	Increment index.
K ≤ N ⟳ Goto 1	K ≤ N ⟳ Goto 1	If K≤N, repeat from Lbl 1.
"SUM":S	"SUM":S	Display value of partial sum.

OPERATION

Preparation

To use this program, store the formula for the general term a_k of the series in program $\emptyset$ (CASIO 7000) or function f_1 (CASIO 7700) as a function of K. For example, to find partial sums of the series $\sum_{k=1}^{\infty} \frac{1}{k}$, program $\emptyset$ or function f_1 must be entered as $1/K$.

Input

Once the formula for a_k is stored in program $\emptyset$ or function f_1, then to find the partial sum $\sum_{k=I}^{N} a_k$, input I, the initial value of the index, and N, the final value of the index.

Output

The program will output the value of the partial sum $\sum_{k=I}^{N} a_k$.

Sample Run

For the series $\sum_{k=1}^{\infty} \frac{1}{k}$, enter program $\emptyset$ (CASIO 7000) or function f_1 (CASIO 7700) as 1/K. Run program SERIES and input I = 1 and N = 10. The partial sum is $\sum_{k=1}^{10} \frac{1}{k}$ = 2.928968254. Run the program again using I = 4 and N = 20. The partial sum is $\sum_{k=4}^{20} \frac{1}{k}$ = 1.764406324.

A.20 Spider

Program SPIDER sketches the graph of a distance-versus-time function while displaying the actual movement of the object (spider?) in real time as it climbs vertically on the right side of the screen. The object has the same vertical movement as the y coordinate on the distance-versus-time graph. The movement of the object is isolated so that the rate at which the object moves is prominent. This program requires 94 (CASIO 7000) or 86 (CASIO 7700) bytes of memory. SPIDER is useful in providing insight into rate of change for graphs of **monotonic** functions only.

CODE: SPIDER

Code (CASIO 7000)	Code (CASIO 7700)	Comments
"SPIDER"	SPIDER	Program title.
Cls	Cls	Clear the graphics screen.
"A"? $\rightarrow$ A	"A"? $\rightarrow$ A	Input Xmin. Store in A.
"B"? $\rightarrow$ B	"B"? $\rightarrow$ B	Input Xmax. Store in B.
(B – A) ÷ 64 $\rightarrow$ S	(B – A) ÷ 64 $\rightarrow$ S	Determine step size.
Range A, B, 1	Range A, B, 1	Set x-interval in Range.
A + .1S $\rightarrow$ T	A + .1S $\rightarrow$ T	Initialize T slightly larger than A.
Prog 2	Prog 2	Call program Cls-TEXT as subroutine.

(continued)

Code (CASIO 7000)	Code (CASIO 7700)	Comments
Lbl 1	Lbl 1	First place to Goto.
Prog ∅:Ans → Y		Determine f(T). Store in Y.
Plot T, Y	Plot T, f_1	Plot the point (T, f(T)).
Plot B – 2S, Y	Plot B – 2S, f_1	Plot spider's position.
T + S → T	T + S → T	Increment T.
T < (B – 2S) ⇨ Goto 1	T < (B – 2S) ⇨ Goto 1	Test for termination of graph.
Plot	Plot	Keep graph visible.

OPERATION

Preparation

Set the range to display the graph in an appropriate viewing rectangle. The domain is set from the program. Store program Cls-TEXT in program 2 (not necessary for *Fx*—7000 GA).

<u>CASIO 7000</u> Store the formula for the monotonic function to be displayed in program ∅ as a function of T. For example, to sketch the distance versus time graph for a spider moving at the rate of $f(x) = \ln x$, store program ∅ as ln T.

<u>CASIO 7700</u> Enter the formula for the monotonic function to be displayed in function f_1 as a function of T. For example, to sketch the distance versus time graph for a spider moving at the rate of $f(x) = \ln x$, store function f_1 as ln T.

Input

Input the values of Xmin and Xmax for the viewing rectangle at the prompts **A?** and **B?** respectively.

Output

The graph of the function in program ∅ (CASIO 7000) or f_1 (CASIO 7700) is sketched as an object moves vertically on the right side of the screen. The object has the same vertical movement as points on the graph of f.

Sample Run

To graph $f(x) = \ln x$ and observe the rate at which f changes, store the formula ln T in program ∅ (CASIO 7000) or function f_1 (CASIO 7700). Set the y-interval for the viewing rectangle to [–2, 2]. Run program SPIDER and watch the object move on the right side of the screen as the function $y = f(x)$ is being drawn. Notice where the object slows, stops, speeds up, etc.

A.21 Square Graphics Screen

Program SQ.SCREEN takes the values of x and y intervals provided by the user, tests to determine which interval is proportionately larger, and sets the Range to allow for the largest interval, while making the x and y scales the same. SQ.SCREEN centers the viewing rectangle about the midpoint of the x and y intervals chosen by the user. This *square* viewing rectangle allows the graphs of conics (and functions) to appear with the proper dimensions. Circles appear as circles and ellipses appear as ellipses. This program requires 131 (CASIO 7000) or 130 (CASIO 7700) bytes of memory.

CODE: SQ.SCREEN

Code (CASIO 7000)	Code (CASIO 7700)	Comments
"SQ.SCREEN"	SQ.SCREEN	Title of program.
"A"? $\rightarrow$ A	"A"? $\rightarrow$ A	Input Xmin. Store in A.
"B"? $\rightarrow$ B	"B"? $\rightarrow$ B	Input Xmax. Store in B.
"C"? $\rightarrow$ C	"C"? $\rightarrow$ C	Input Ymin. Store in C.
"D"? $\rightarrow$ D	"D"? $\rightarrow$ D	Input Ymax. Store in D.
(B − A) ÷ 9.4 $\rightarrow$ E	(B − A) ÷ 9.4 $\rightarrow$ E	Find factor for x.
(D − C) ÷ 6.2 $\rightarrow$ F	(D − C) ÷ 6.2 $\rightarrow$ F	Find factor for y.
F < E ⇨ E $\rightarrow$ F	F < E ⇨ E $\rightarrow$ F	If x-factor larger, change factor.
(A + B) ÷ 2 $\rightarrow$ P	(A + B) ÷ 2 $\rightarrow$ P	Find x-interval midpoint.
(C + D) ÷ 2 $\rightarrow$ Q	(C + D) ÷ 2 $\rightarrow$ Q	Find y-interval midpoint.
Int F + 1 $\rightarrow$ S	Int F + 1 $\rightarrow$ S	Store step size.
Range P − 4.7F, P + 4.7F, S, Q − 3.1F, Q + 3.1F, S	Range P − 4.7F, P + 4.7F, S, Q − 3.1F, Q + 3.1F, S	Set Range centered on x and y midpoints.

OPERATION

There is no preparation to be done. SQ.SCREEN is self-contained.

Input

Enter the values of Xmin, Xmax, Ymin, and Ymax at the prompts **A?**, **B?**, **C?**, and **D?**, respectively. SQ.SCREEN adjusts the Range so that x and y intervals are shown with the same scales. SQ.SCREEN can be used to set the Range for a particular x or y interval centered about a particular choice of x or y as follows: If a particular x-interval is desired, enter the values for the endpoints of that x-interval. Enter the value of y that is to serve as the center for the y-interval for both C and D. Similarly, if a particular y-interval is desired, enter the value for the center of the x-interval into both A and B followed by the desired values of C and D.

Output

When SQ.SCREEN is run, the Range is changed. SQ.SCREEN alters the smaller interval so that the scales for x and y are the same. SQ.SCREEN also sets the tick marks for x and y to be the same distance apart.

Sample Run

To set the Range for the x-interval [–3, 3] and centered about the x-axis, run SQ.SCREEN with A = –3, B = 3, C = 0, and D = 0. Press the [Range] key to display the values of Xmin = –3, Xmax = 3, Ymin = –1.9787234, Ymax = 1.9787234, Xscl = Yscl = 1. To set the Range for the y-interval [–10, 10], centered about the line x = 3, run SQ.SCREEN with A = 3, B = 3, C = –10, and D = 10. Press the [Range] key to display the values of Xmin = –12.1612903, Xmax = 18.1612903, Ymin = –10, Ymax = 10, Xscl = Yscl = 4.

A.22 Tangent

Program TANGENT sketches the graph of a function $y = f(x)$ and then draws tangent lines whose slopes are approximated by the values of

$$m = \frac{f(x + .01) - f(x - .01)}{.02}$$

for each value of x that is used. In addition, as the tangent lines are drawn, the points (x, m) are plotted, leaving a trace of the graph of the derivative function on the screen. This program uses 128 (CASIO 7000) or 124 (CASIO 7700) bytes of memory. TANGENT calls program NDERIV (subroutine code) as a subroutine. The subroutine requires 38 (CASIO 7000) or 40 (CASIO 7700) bytes of memory.

CODE: TANGENT

Code (CASIO 7000)	Code (CASIO 7700)	Comments
"TANGENT"	TANGENT	Title of program.
Cls	Cls	Clear graphics screen.
"A"? → A	"A"? → A	Input Xmin. Store in A.
"B"? → B	"B"? → B	Input Xmax. Store in B.

(*continued*)

Code (CASIO 7000)	Code (CASIO 7700)	Comments
Range A, B, 1	Range A, B, 1	Set x-values in Range.
$(B - A) \div 30 \rightarrow S$	$(B - A) \div 30 \rightarrow S$	Determine step size.
$3S \rightarrow C$	$3S \rightarrow C$	Determine length of tangent segments.
Prog 1◢	Graph Y = f_1◢	Graph $y = f(x)$.
$A + C \rightarrow P$	$A + C \rightarrow P$	Assign value of $A + C$ to P.
Lbl 1	Lbl 1	First Lbl to Goto.
$P \rightarrow X$	$P \rightarrow X$	Store value of P in X.
Prog ∅:Ans $\rightarrow$ Q	$f_1 \rightarrow Q$	Store $f(P)$ in Q
$.01 \rightarrow H$	$.01 \rightarrow H$	Initialize H.
Prog 3	Prog 3	Call program NDERIV as subroutine.
Plot P, M	Plot P, M	Plot point $(P, f\,'(P))$.
Plot P – C, Q – MC	Plot P – C, Q – MC	Plot endpoint of tangent segment.
Plot P + C, Q + MC	Plot P + C, Q + MC	Plot endpoint of tangent segment.
Line◢	Line◢	Draw segment.
$P + S \rightarrow P$	$P + S \rightarrow P$	Increment P.
$P + C \leq B$ ⇨ Goto 1	$P + C \leq B$ ⇨ Goto 1	If still in viewing rectangle, repeat from Lbl 1.
"END"	"END"	Print END if done.

OPERATION

Preparation

Set the range to display the graph in an appropriate viewing rectangle. Enter the subroutine code for program NDERIV into program 3. Note: If NDERIV is entered into a different numbered program location, be sure to enter that program number into the code for TANGENT in place of the code **Prog 3**.

CASIO 7000

Enter the formula for the function to be displayed in program ∅. Enter the graph statement into program 1. For example, to sketch tangents along the graph of $f(x) = x^2$, store program ∅ as X^2. Store program 1 as Graph Y = X^2.

CASIO 7700

Enter the formula for the function to be displayed in function f_1.

Input

Input the values of Xmin and Xmax for the viewing rectangle at the prompts **A?** and **B?**, respectively.

Output

The program first sketches the graph of the function stored in programs Ø and 1 (CASIO 7000) or f_1 (CASIO 7700). Each time the | EXE | key is pressed, the program draws the line tangent to the graph of f at $(x, f(x))$ and it plots the point (x, m) where m is the slope of the tangent line. It repeats this process for about 25 different values of x. When the program is completed, it will display the word END. To view the completed graph, press the | G↔T | key.

Sample Run

Preparation

To sketch the graph of $f(x) = x^3 - 3x$ and its derivative, set the Range to view the graphs of the rectangle [–2.5, 2.5] by [–7, 7].

> **CASIO 7000**

Enter the formula for the function into program Ø as $X^3 - 3X$. Enter the graph statement into program 1 as Graph $Y = X^3 - 3X$.

> **CASIO 7700**

Enter the formula for the function into f_1 as $X^3 - 3X$.

Run

Run program TANGENT. Press the | EXE | key several times to draw tangent lines along the curve of f and to plot the points on the graph of the derivative function. The points on the graph of the derivative function should resemble the graph of a parabola. When the program is completed, press the | G↔T | key to view the graphics screen again.

A.23 Vectors

Programs VECTORS.2D and VECTORS.3D are used to compute the sum, difference, and dot product of two vectors, and to compute the norm of the sum and difference of the vectors. In addition, VECTORS.2D illustrates the sum and the difference of vectors, and VECTORS.3D computes the cross product for three-dimensional vectors. The code for VECTORS.2D is written so that it is easily revised to accomodate the code for VECTORS.3D.

CODE: VECTORS.2D

VECTORS.2D sketches vectors v_1 and v_1, asks the user to determine which operation is to be performed, performs and illustrates the operation, then sketches the resultant vector and displays the norm. VECTORS.2D requires 252 (CASIO 7000) or 251 (CASIO 7700) bytes of memory.

Code (CASIO 7000)	Code (CASIO 7700)	Comments
"VECTORS.2D"	VECTORS.2D	Title of program.
"V1"? → A	"V1"? → A	Prompt for vector v_1, input x component in A.
? → B	? → B	Input y component in B.
"V2"? → D	"V2"? → D	Prompt for vector v_2, input x component in D.
? → E	? → E	Input y component in E.
Lbl 7	Lbl 7	Seventh label to Goto.
Cls	Cls	Clear graphics screen.
Plot ∅, ∅:Plot A, B:Line◢	Plot ∅, ∅:Plot A, B:Line◢	Draw vector v_1.
Plot ∅, ∅:Plot D, E:Line◢	Plot ∅, ∅:Plot D, E:Line◢	Draw vector v_2.
∅ → P	∅ → P	Initialize operation variable.
"OPERATION A(1), S(2), D(3)"? → P	"OPERATION A(1), S(2), D(3)"? → P	Prompt for operation input, store 1, 2, or 3 in P.
P = 1 ⇨ Goto 1	P = 1 ⇨ Goto 1	If P = 1, Goto 1.
P = 2 ⇨ Goto 2	P = 2 ⇨ Goto 2	If P = 2, Goto 2.
"DP": AD + BE◢	"DP": AD + BE◢	Compute dot product.
Goto 6	Goto 6	Send control to Lbl 6.
Lbl 1	Lbl 1	First label to Goto.
A + D → I	A + D → I	Add x components.
B + E → J	B + E → J	Add y components.
Goto 5	Goto 5	Send control to Lbl 5.
Lbl 2	Lbl 2	Second Lbl to Goto.
A − D → I	A − D → I	Subtract x components.
B − E → J	B − E → J	Subtract y components.
Lbl 5	Lbl 5	Fifth Lbl to Goto.
Plot A, B:Plot I, J:Line◢	Plot A, B:Plot I, J:Line◢	Plot vector $v_1 \pm v_2$.
Plot ∅, ∅:Plot I, J:Line◢	Plot ∅, ∅:Plot I, J:Line◢	Plot resultant vector.

(continued)

Code (CASIO 7000)	Code (CASIO 7700)	Comments
"I J": I◢	"I J": I◢	Print x component of resultant vector.
J◢	J◢	Print y component of resultant vector.
"NORM": $\sqrt{(I^2 + J^2)}$◢	"NORM": $\sqrt{(I^2 + J^2)}$◢	Print NORM and value.
Lbl 6	Lbl 6	Sixth Lbl to Goto.
"SAME VECTORS Y(1), N(2)"? → Z	"SAME VECTORS Y(1), N(2)"? → Z	Prompt for repeat with same or new vectors.
Z = 1 ⇨ Goto 7	Z = 1 ⇨ Goto 7	If same vectors, Goto 7.

OPERATION: VECTORS.2D

Preparation

Prepare the Range to sketch the vectors on a viewing rectangle in which both v_1, v_2 and the resultant vector $v_1 + v_2$ or $v_1 - v_2$ can be shown. If scalar multiples are needed for any of the vectors, determine these before entering them into the program, and adjust the Range accordingly.

Input

Run program VECTORS.2D. Enter the first component of v_1 after the prompt **V1?**. Enter the second component for v_1 after **?**. Enter the first component of v_2 after the prompt **V2?**. Enter the second component for v_2 after **?**. Press $\boxed{\text{EXE}}$ twice to display vectors v_1 and v_2. Choose the operation **A**ddition (1), **S**ubtraction (2), or **D**ot Product (3). Once the operation has been completed, choose whether or not the SAME Y(1) or N(2) vectors are to be used in another operation.

Output

Once the vectors are entered, the vectors v_1 and v_2 are displayed with common originating point of (0, 0). The operation is chosen and the result of the operation is illustrated and displayed. The norm of the resultant vector is also displayed.

Sample Run

To illustrate and display the sum and difference of vectors $v_1 = <2, 4>$ and $v_2 = <3, -1>$, set the Range to [−9.4, 9.4] by [−6.2, 6.2]. Run program VECTORS.2D. Enter the x and y components of v_1 after the **V1?** and **?** prompts respectively. Enter the x and y components of v_2 after the **V2?** and **?** prompts respectively. Press $\boxed{\text{EXE}}$ after each input. Press $\boxed{\text{EXE}}$ to sketch v_1, then once again to sketch v_2. Press ⌈EXE⌉ to return to the text screen and choose the desired operation by pressing 1, 2, or 3 for **A**ddition, **S**ubtraction, or **D**ot product respectively. After you press 1, v_2 is sketched with its originating point at the end of v_1. Press ⌈EXE⌉ to display the sketch of the resultant vector from the origin to the terminating point of $v_1 + v_2$. Press ⌈EXE⌉ three more times to display the resultant vector <I, J> = <5, 3> and NORM = 5.830951895. Run the program again, choosing 1 for SAME vectors, then 2 for subtraction. Notice the similarities and differences in the geometric representation of the sum and difference of these two vectors. This time <I, J> = <-1, 5> and NORM = 5.099019514. To compute the dot product, press 1 to use the same vectors once more, then choose operation 3. DP = 2.

Adaptation

VECTORS.2D can be made more general to illustrate the sum of any two vectors with originating point at (P, Q) by replacing all lines **Plot Ø, Ø** in the program to **Plot P, Q**. Also, all of the other plot statements must be changed to add P to the *x* component and Q to the *y* component. If you would like VECTORS.2D to determine and use scalar multiples of vectors v_1 and v_2, after the input lines for v_1, insert the code: **"S"? → S: AS → A: BS → B** After the input lines for v_2, insert the code: **"R"? → R: DR → D: ER → E** The rest of the program remains the same and operates as above except for a request for input of the scalar S for v_1 and scalar R for v_2.

VECTORS.3D

VECTORS.3D computes sums, differences, norms of sums and differences, dot products and cross products of three-dimensional vectors. The code for VECTORS.3D is a revision of the code for VECTORS.2D. Notice that many of the lines of code remain the same. All of the graphic commands have been removed. Additional code includes code for the z components for v_1 and v_2 and code for the cross product calculations have been added. VECTORS.3D requires 274 (CASIO 7000) or 273 (CASIO 7700) bytes of memory. A shortened form of VECTORS.3D that can be used to compute dot products, cross products, and norms is recommended for CASIO 7000 users. It is provided in the adaptations on page 214.

Code (CASIO 7000)	Code (CASIO 7700)	Comments
"VECTORS.3D"	VECTORS.3D	Title of program.
"V1"? → A	"V1"? → A	Prompt for vector v_1, input *x* component in A.
? → B	? → B	Input *y* component in B.

(continued)

Code (CASIO 7000)	Code (CASIO 7700)	Comments
$? \to C$	$? \to C$	Input z component in C.
"V2"? $\to D$	"V2"? $\to D$	Prompt for vector v_2, input x component in D.
$? \to E$	$? \to E$	Input y component in E.
$? \to F$	$? \to F$	Input z component in F.
Lbl 7	Lbl 7	Seventh label to Goto.
$\emptyset \to P$	$\emptyset \to P$	Initialize operation variable.
"OPERATION A(1), S(2), D(3), C(4)"? $\to P$	"OPERATION A(1), S(2), D(3), C(4)"? $\to P$	Prompt for operation input, store 1, 2, 3, or 4 in P.
$P = 1 \diamond$ Goto 1	$P = 1 \diamond$ Goto 1	If P = 1, Goto 1.
$P = 2 \diamond$ Goto 2	$P = 2 \diamond$ Goto 2	If P = 2, Goto 2.
$P = 4 \diamond$ Goto 4	$P = 4 \diamond$ Goto 4	If P = 4, Goto 4.
"DP": AD + BE + CF◢	"DP": AD + BE + CF◢	Compute dot product.
Goto 6	Goto 6	Send control to Lbl 6.
Lbl 1	Lbl 1	First label to Goto.
$A + D \to I$	$A + D \to I$	Add x components.
$B + E \to J$	$B + E \to J$	Add y components.
$C + F \to K$	$C + F \to K$	Add z components.
Goto 5	Goto 5	Send control to Lbl 5.
Lbl 2	Lbl 2	Second Lbl to Goto.
$A - D \to I$	$A - D \to I$	Subtract x components.
$B - E \to J$	$B - E \to J$	Subtract y components.
$C - F \to K$	$C - F \to K$	Subtract z components.
Goto 5	Goto 5	Send control to Lbl 5.
Lbl 4	Lbl 4	Fourth Lbl to Goto.
$BF - CE \to I$	$BF - CE \to I$	Find x component of cross product.
$CD - AF \to J$	$CD - AF \to J$	Find y component of cross product.
$AE - BD \to K$	$AE - BD \to K$	Find z component of cross product.
Lbl 5	Lbl 5	Fifth Lbl to Goto.
"I J K": I◢	"I J K": I◢	Print x component of resultant vector.

(continued)

Code (CASIO 7000)	Code (CASIO 7700)	Comments
J◢	J◢	Print y component of resultant vector.
K◢	K◢	Print z component of resultant vector.
"NORM": $\sqrt{(I^2 + J^2 + K^2)}$◢	"NORM": $\sqrt{(I^2 + J^2 + K^2)}$◢	Print NORM and value.
Lbl 6	Lbl 6	Sixth Lbl to Goto.
"SAME VECTORS Y(1), N(2)"? → Z	"SAME VECTORS Y(1), N(2)"? → Z	Prompt for repeat with same or new vectors.
Z = 1 ⇨ Goto 7	Z = 1 ⇨ Goto 7	If same vectors, Goto 7.

OPERATION: VECTORS.3D

Preparation

If scalar multiples are needed for any of the vectors, determine these before entering the vectors into the program.

Input

Run program VECTORS.3D. Enter the first component of v_1 after the prompt **V1?**. Enter the second component for v_1 after ?, and the third component for v_1 after ?. Enter the first component of v_2 after the prompt **V2?**. Enter the second and third components for v_2 after each successive ?. Press $\boxed{\text{EXE}}$. Choose the operation **A**ddition (1), **S**ubtraction (2), **D**ot Product (3), or **C**ross product (4). Once the operation has been completed, choose whether or not the SAME Y(1) or N(2) vectors are to be used in another operation.

Output

Once the vectors are entered, depending on the operation chosen, the resultant vector <I, J, K> is displayed by pressing $\boxed{\text{EXE}}$ three times, or the dot product (indicated by DP) is displayed. The norm of the resultant vector is also displayed for the sum, difference, or cross product vectors.

Sample Run

To display the sum, difference, dot product, and cross product of vectors $v_1 = <1, 2, 1>$ and $v_2 = <1, 3, 2>$, run program VECTORS.3D. Enter the x, y, and z components of v_1 after the **V1?** and ? prompts, respectively. Enter the x, y, and z components of v_2 after the **V2?** and ? prompts, respectively. Press $\boxed{\text{EXE}}$ after each input. Choose the desired operation by pressing 1, 2, 3, or 4 for **A**ddition, **S**ubtraction, **D**ot product, or **C**ross product respectively. After pressing 1, Press

$\boxed{\text{EXE}}$ three more times to display the resultant vector <I, J, K> = <2, 5, 3> and NORM = 6.164414003. Run the program again choosing 1 for same vectors, then 2 for subtraction. This time <I, J, K> = <0, −1, −1> and NORM = 1.414213562. To compute the dot product, press 1 to use the same vectors once more, then choose operation 3. DP = 9. To compute the cross product, press 1 to use the same vectors once more, then choose operation 4. <I, J, K> = <1, −1, 1> with NORM of 1.732050808.

Adaptations

An adaptation of VECTORS.3D that allows scalar multiplication for v_1 and v_2 is the same as for VECTORS.2D. VECTORS.3D is adapted below to compute only the dot product, the cross product, and the norm of the cross product. V.3D can be used in place of VECTORS.3D. Vector addition, subtraction, and norms of resultants must be computed with paper and pencil. This version of VECTORS.3D requires 109 bytes of memory.

Code	Comments
"V .3D"	Title of program.
"V1"? → A	Prompt for vector v_1, input x component in A.
? → B	Input y component in B.
? → C	Input z component in C.
"V2"? → D	Prompt for vector v_2, input x component in D.
? → E	Input y component in E.
? → F	Input z component in F.
"DP": AD + BE + CF◢	Compute dot product.
"CP, <I J K>"	
BF − CE → I◢	Find x component of cross product.
CD − AF → J◢	Find y component of cross product.
AE − BD → K◢	Find z component of cross product.
"NORM": $\sqrt{(I^2 + J^2 + K^2)}$	Print NORM and value.

No preparation is required to run V.3D. Input is the same as for VECTORS.3D except that it is no longer necessary to indicate the operation and use of same vectors. V.3D outputs the dot product (DP), the cross product (CP), and the norm of the cross product.

A.24 Zoom

The program ZOOM allows the user to zoom in on a particular portion of the graph of a function (or the graphs of functions) by capturing that portion by a rectangular window called the viewing rectangle. The figure on page 215 shows how the

CASIO screen appears when the user is ready to zoom in on that area of the screen contained in the small viewing rectangle. ZOOM is included for CASIO 7000 users. This option is already available on the CASIO 7700. The program requires 127 bytes of memory. ZOOM was written by Dr. Charles Vonder Embse, Department of Mathematics, Central Michigan University.

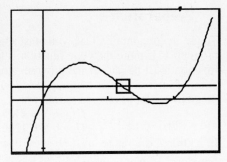

CODE: ZOOM

Code (CASIO 7000)	Comments
"ZOOM"	Title of the program.
Cls	Clear the graphics screen.
Lbl 1	First label to Goto.
Prog 1	Graph the function $y = f(x)$.
Plot◢	Put cursor in middle of screen and pause.
X → A:Y → D	Store coordinates of selected point into A and D.
Plot◢	Second cursor, pause until EXE is pressed.
X → B:Y → C	Store coordinates of selected point into B and C.
B>A ⇨ Goto 2	Check order. If B > A then Goto 2.
A → T:B → A:T → B	If B < A, this code swaps A and B.
Lbl 2	Second label to Goto.
D>C ⇨ Goto 3	Check order; if D > C then Goto 3.
C → T:D → C:T → D	If D < C, this code swaps C and D.
Lbl 3	Third label to Goto.
Plot A, D	Plot (A, D), upper left corner of rectangle.
Plot B, D : Line	Plot (B, D), upper right corner. Draw top line.
Plot B, C : Line	Plot (B, C), lower right corner. Draw right line.
Plot A, C : Line	Plot (A, C), lower left corner. Draw bottom line.
Plot A, D : Line◢	Replot (A, D). Finish drawing rectangle. Pause.
Range A, B, 1, C, D, 1	Set new range.
Goto 1	Goto 1 to regraph function(s) in the new range and repeat process.

OPERATION

Preparation

To use the program ZOOM, you must store a graph statement for the function(s) in program 1. If more than one function is to be graphed, a graph statement for each function must be part of the program stored in program 1. For example, to find a point of intersection of the graphs of $f(x) = 4x^3 - 140x^2 + 1200x$ and $g(x) = 1200$, program 1 must be entered as follows:

$$\text{Graph Y} = 4X^3 - 140\ X^2 + 1200X$$
$$\text{Graph Y} = 1200$$

In addition, the range must be set properly to obtain an appropriate graph of the functions. It may be necessary to use the $\boxed{\text{Range}}$ key and run program 1 several times in order to get the range set appropriately.

Input

The input for this program is quite different than most programs. The diagonal corners of the small viewing rectangle must be defined. When the program is run, a graph of the function(s) will appear on the screen. When the graph is finished, a flashing point will appear in the center of the screen with the x-coordinate of the point at the bottom of the screen. (To see the y-coordinate, press the $\boxed{\text{X}\leftrightarrow\text{Y}}$ key. Press this key again to return to the x-coordinate.) Move the flashing point using the four cursor control keys to one corner of the desired viewing rectangle. When the flashing point is at one of the corners of the desired viewing rectangle, press the $\boxed{\text{EXE}}$ key to set that corner of the viewing rectangle. Another flashing point will appear in the middle of the screen. Move this second flashing point to the corner of the viewing rectangle that is diagonally opposite the corner that you previously set. Once this second flashing point is in position, press the $\boxed{\text{EXE}}$ key to set that corner of the viewing rectangle.

Output

Once the diagonal corners of the zoom rectangle are defined, the viewing rectangle is drawn on the screen. Press the $\boxed{\text{EXE}}$ key again. The graph(s) of the function(s) are drawn and the viewing rectangle is blown up like a photograph to fill the entire screen. You may check the range of the viewing screen at any time by pressing the $\boxed{\text{Range}}$ key. (Pressing the $\boxed{\text{Range}}$ key a second time returns the calculator to the graphics screen.)

This procedure can be repeated as many times as necessary to get a good approximation of a point of interest on the graph(s).

Sample Run

<u>Preparation</u>

To use the program ZOOM to find a point of intersection of the graphs of $f(x) = 4x^3 - 140x^2 + 1200x$ and $g(x) = 1200$, enter program 1 as follows:

Graph Y = 4X^3 – 140X^2 + 1200X

Graph Y = 1200

Set the Range to view these graphs on the viewing rectangle [–5, 30] by [–1000, 3500].

<u>Run</u>

Run program ZOOM. The graphs on the screen should show three points of intersection for the two graphs. The x-coordinate of the flashing point in the middle of the screen X = 12.5 appears on the bottom of the screen. (The y-coordinate should be 1250.) To approximate the coordinates of the middle point of intersection, move the flashing point just to the left and above the point of intersection and press the $\boxed{\text{EXE}}$ key. Move the second flashing point just to the right and below the point of intersection and press $\boxed{\text{EXE}}$. This produces a viewing rectangle that contains that point of intersection. Press $\boxed{\text{EXE}}$ again and the program magnifies the viewing rectangle to fill the entire screen.

By zooming in several times and checking the range, you can determine the coordinates for the point of intersection. Correct to the nearest hundredth, the x-coordinate of the point of intersection is 11.89. The y-coordinate of the point of intersection is 1200.

Adaptation

ZOOM is a program that is used quite often in *Exploring Calculus*. It is helpful to keep this program stored in the CASIO 7000 series calculator at all times. Since the memory of the CASIO 7000 is particularly limited, it is useful to keep the number of bytes used in this program to a minimum.

The following lines can be deleted from ZOOM without changing the results provided by ZOOM if the opposite corners of the ZOOM rectangle are chosen as follows: Position the first blinking cursor at the upper left corner of the ZOOM box. Position the second blinking cursor at the lower right corner of the ZOOM box. This eliminates the need to swap the values of A and B and/or C and D.

B>A ⇨ Goto 2

A → T:B → A:T → B

Lbl 2

D>C ⇨ Goto 3

C → T:D → C:T → D

Lbl 3

This adaptation requires only 83 bytes of memory. (Note: Removing the title also decreases the amount of required memory.)

B

Texas Instrument
TI-81 Appendix

B.1 TI-81 Programs for *Exploring Calculus*

Listed in the first column are the titles of the explorations in this collection. The second column lists the programs necessary to complete each of the explorations. The third column lists the pages on which the necessary programs can be found.

Exploration	Program(s)	Page(s)
1. Exploring Families of Functions	FAMILY	234
2. Extended Families of Functions	FAMILY (optional)	
3. Understanding Function Notation	None	
4. Behavior at a Point and End Behavior	BEHAVIOR	230
5. Continuity		
6. Understanding the Derivative	SPIDER SECANT	257
7. Hierarchy of Functions	None	254
8. Graphical Differentiation	TANGENT	258
9. Max/Min Problems	None	
10. Linear Approximations	LINAPPRX (In footnote)	52
	NEWTON-G	240
	NEWTON	242
11. Interpreting the Second Derivative	SPIDER	257
	TANGENT	258
12. A Development of the Function $F(x) = e^x$	SECANT or	253
	BEHAVIOR	230

B.2 Using the TI-81 Graphics Calculator with *Exploring Calculus*

To make best use of these materials, the following suggestions are provided.

CONVENTIONS

- Store functions to be graphed and/or evaluated in function Y_1 (found in the $\boxed{Y=}$ menu). Programs that graph or evaluate functions have all been written to use function Y_1 in the program.
- When entering functions such as $y = \sin x$ into the calculator, use the designated X key, $\boxed{X|T}$.
- To leave a menu, press $\boxed{2nd}$ $\boxed{CLEAR}$ (to quit the menu), or press another menu key.
- To graph functions such as $y = |x|$ and $y = \sqrt{4 - x^2}$ (semi-circle) so that these appear on the screen as expected, use the viewing rectangle [–4.8, 4.7] by [–3.2, 3.1], or a multiple of this viewing rectangle. This viewing rectangle also allows the trace cursor to be placed directly on the origin. To obtain this rectangle, press $\boxed{RANGE}$ to view the range screen and enter each value appropriately.
- In the calculus explorations, when the suggested viewing rectangle is given as [–5, 5] by [–3, 3], the viewing rectangle [–4.8, 4.7] by [–3.2, 3.1] is suitable.

Since this is not one of the built-in ranges, it is advisable to enter program RANGE into program 1. Program RANGE can be found on page 248.

- In the range, the scale settings determine the tick marks on the *x* and *y* axes. Scale markings are not suggested in the materials. Choosing a scale that will allow about ten tick marks to be shown will help you interpret graphs.

- To overlay several functions, enter up to four functions into the $\boxed{Y=}$ menu. Highlight the equal sign for each function by moving the cursor over the equal sign and pressing $\boxed{ENTER}$ for each function to be graphed. Other functions can also be graphed by using the DrawF option in the DRAW menu. Press $\boxed{2nd}$ $\boxed{PRGM}$ followed by the expression to be graphed.

- To interrupt a graph as it is being drawn, press $\boxed{ON}$.

- To terminate a program, press $\boxed{ON}$ or $\boxed{2nd}$ $\boxed{CLEAR}$ (for QUIT).

- The cursor keys operate similarly to cursor control keys on a computer. In a program, these can be used to move the cursor to any position in the text to edit the program.

- The insert and delete keys operate similarly to those on a computer. Pressing $\boxed{INS}$ will allow you to insert characters in the location immediately preceding the highlighted position. Pressing $\boxed{DEL}$ when a character is highlighted will delete the highlighted character.

- Current calculator modes can be checked or changed from the $\boxed{MODE}$ menu. Mode settings include normal, scientific, or engineering notation; floating decimal point; radian or degree mode; function or parametric grapher; point or connected graph plotting; sequential or simultaneous graphing; plotting on grid or without grid; and rectangular or polar coordinate display.

- None of the programs included in this appendix have used the letter O (oh) as a variable. All symbols 0 are zeros.

PREVENTING OR CORRECTING COMMON PROBLEMS

- Check the mode in the calculator whenever an unexpected answer or graph appears. The calculator should be set to radian mode for these materials. Graphs of the trigonometric functions do not appear on the screen for the viewing rectangles suggested if the calculator is set for degree mode.

- If vertical lines appear when you are graphing functions with vertical asymptotes, change the mode setting to point plotting and redraw the graph.

- When the error screen appears while you are inserting programs and expressions to be evaluated, choose option **1:Go to error**. The cursor will appear at a point in the expression or program where the error was committed.

- CAUTION! Take care when using the RESET key. This is the menu from which all memory in the calculator can be inadvertently erased.

B.3 TI-81 Reference Sheet I

Basic Keys	To do:
ON	Turns on the calculator, interrupts graph as it is being sketched, or terminates a program.
X|T	Designated X variable for entering functions.
CLEAR	Clears the text screen.
ENTER	Executes a command; works like the $=$ key on a conventional calculator.
2nd	Press to access functions and menus printed in blue above the keys.
ALPHA	Press to access characters printed in gray above the keys.
2nd ON	OFF: Turns calculator off.
2nd CLEAR	QUIT: Terminates programs, allows user to exit menus.
2nd ENTER	ENTRY: Allows last executed line to be edited and re-executed.
(–)	Negative sign (do not use the subtraction key to indicate negative numbers).
2nd (–)	ANS: Returns the value of the last executed computation.
MODE	Highlighted values display current operating mode of calculator.
Prog n	Executes program number n.
2nd ^	Gives stored estimate of pi.
INS	Allows text to be inserted before the character highlighted by the cursor.
DEL	Allows text highlighted by the cursor to be deleted.
n STO▶ A	Store the value n in memory location A. (No need to use ALPHA key to enter variable name.)
MATH	MATH menu has more functions (see TI-Reference Sheet II).
▲	Recall last command entered interactively.

Graphics Menus and Commands

Basic Operations	To do:
Y=	Y= menu: Stores up to four functions for use in function evaluation and graphing. When equal sign is highlighted, function is active for graphing.

Basic Operations	To do:
RANGE	Allows user to change domain, range, and scale values, and change number of points plotted.
ZOOM	Zoom menu allows user to zoom in, zoom out, and choose square, standard, trigonometric, and integer windows for graphic displays..
TRACE	Allows user to determine x and y coordinates of points along one or more active graphs.
GRAPH	Graphs function(s) highlighted in Y= menu, toggles between graphics and text screens.
DRAW 2nd PRGM	DRAW menu: Allows user to clear the graphics screen, plot lines, point, other functions, and shade between two functions.
Clr Draw 2nd PRGM 1	Clears the graphics screen.
Line (2nd PRGM 2 x_1, y_1, x_2, y_2	Draws a line between the two points (x_1, y_1), (x_2, y_2).
PT-On (2nd PRGM 3 m, n	Plots the point (m, n) where m and n are two numbers in the current calculator range. When graphics screen is active, PT-On centers the cursor on the graphics screen.
DrawF 2nd PRGM 6 <expression>	Allows user to sketch the graph of the function $f(x)$ = <expression>.
Shade (2nd PRGM 7	Allows user to shade between two functions entered in the Y= menu.

B.4 TI-81 Reference Sheet II

The TI-81 graphics calculator has been developed to operate from menus. This reference sheet was created to help you locate the commands and operations that are found in the calculator menus. The calculator keys are listed in boxes. Menus accessed from the calculator keys are typed in capital letters. Only menus used in the calculus materials are included here.

$\boxed{Y=}$	$\boxed{\text{RANGE}}$	$\boxed{\text{ZOOM}}$	$\boxed{\text{MODE}}$
:Y1 =	Xmin =	1:Box	Norm Scl Eng
:Y2 =	Xmax =	2:Zoom In	Float 0123456789
:Y3 =	Xscl =	3:Zoom Out	Rad Deg
:Y4 =	Ymin =	4:Set Factors	Function Param
	Ymax =	5:Square	Connected Dot
	Yscl =	6:Standard	Sequence Simul
	Xres =	7↓Trig	Grid Off Grid On
		8:Integer	Rect Polar

$\boxed{\text{MATH}}$			$\boxed{\text{2nd}}$ $\boxed{\text{MATH}}$
MATH	NUM	HYP	(TEST)
1:R▶P(	1:Round(	1:sinh	1: =
2:P▶R(	2:IPart	2:cosh	2: ≠
3:^{3}	3:FPart	3:tanh	3: >
4:$^3\sqrt{}$	4:Int	4:sinh^{-1}	4: ≥
5:!		5:cosh^{-1}	5: <
6:°		6:tanh^{-1}	6: ≤
7:^{r}			
8:NDeriv(			

$\boxed{\text{PRGM}}$		Pressing	$\boxed{\text{PRGM}}$	while in EDIT	
EXEC/EDIT	ERASE # steps	CTL	I/O	EXEC	
1:Prgm1 NAME	1:Prgm1 18	1:Lbl	1:Disp	1:Prgm1 NAME	
2:Prgm2 NAME	2:Prgm2 24	2:Goto	2:Input	2:Prgm2 NAME	
3:Prgm3 ETC	3:Prgm3 17	3:If	3:DispHome	3:Prgm3 ETC	
4:Prgm4	4:Prgm4	4:IS>(	4:DispGraph	4:Prgm4	
5:Prgm5	5:Prgm5	5:DS<(	5:ClrHome	5:Prgm5	
6:Prgm6	6:Prgm6	6:Pause		6:Prgm6	
7↓Prgm7	7↓Prgm7	7:End		7↓Prgm7	
		8:Stop			

$\boxed{\text{2nd}}$ $\boxed{\text{PRGM}}$	$\boxed{\text{VARS}}$	$\boxed{\text{2nd}}$ $\boxed{\text{VARS}}$	Y-VARS	
(DRAW)	RNG	Y	ON	OFF
1:ClrDraw	1:Xmin	1:Y1	1:All-On	1:All-Off
2:Line (	2:Xmax	2:Y2	2:Y1-On	2:Y1-Off
3:PT-On (	3:Xscl	3:Y3	3:Y2-On	3:Y2-Off
4:PT-Off (	4:Ymin	4:Y4	4:Y3-On	4:Y3-Off
5:PT-Chg (	5:Ymax	5:X1T	5:Y4-On	5:Y4-Off
6:DrawF	6:Yscl	6:Y1T	6:X1T-On	6:X1T-Off
7:Shade (	7:Xres	7:X2T	7:X2T-On	7:X2T-Off
	8:Tmin	8:Y2T	8:X3T-On	8:X3T-Off
	9:Tmax	9:X3T		
	Ø:Tstep	Ø:Y3T		

B.5 Diving Board Problem[1]

When a person stands on a diving board, the amount the board bends at the point where she is standing, y, below its rest position is a cubic function of x, the distance from the built-in end to the point where she is standing on the board. See the drawing below.

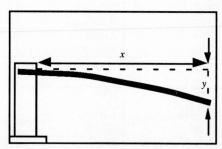

Suppose that the following measurements are taken. Note that the deflection from the horizontal (the bend) is recorded as a negative value.

Horizontal distance in feet from built-in end of board x	Deflection from horizontal in thousandths of an inch y
0	0
2	–528
4	–1784
6	–3576

1. Derive the equation expressing y as a function of x.

    This can be done by writing $y = ax^3 + bx^2 + cx + d$ and then using the four measured values for x and y to determine a system of four equations in four unknowns (a, b, c, and d).

2. What does it mean in this situation for x to be positive? negative? Do each of these make sense? Explain.

3. What does it mean in this situation for y to be positive? negative? Do each of these make sense? Explain.

4. What is an appropriate domain for x? What is an appropriate range? Explain.

    To graph the function on the suggested domain and range, enter these values into the calculator. Press the RANGE key and enter the appropriate values, each followed by ENTER .

5. Sketch the graph of this function.

---

[1] This problem is adapted from Paul Foerster, *Algebra and Trigonometry*, Addison-Wesley, 1990. It is included here to allow students to solve a problem from precalculus while becoming familiar with the TI-81 graphics calculator.

Enter the function to be graphed into the calculator by pressing $\boxed{\text{Y=}}$, then entering the function into Y_1 as follows:

4 $\boxed{\text{X|T}}$ $\boxed{\wedge}$ 3 – 115 $\boxed{\text{X|T}}$ $\boxed{x^2}$ – 50 $\boxed{\text{X|T}}$ $\boxed{\text{GRAPH}}$.

6. Find the zeros of this function and explain what they represent in this situation.

a. To determine the zeros graphically:

Press $\boxed{\text{Trace}}$ and then by pressing the left and right cursor keys, move the blinking cursor to a point close to where the graph crosses the x-axis. The x and y-coordinates of the blinking cursor will be displayed across the bottom of the screen. Press $\boxed{\text{ZOOM}}$ 2 $\boxed{\text{ENTER}}$ to activate the automatic zoom-in.

Press $\boxed{\text{RANGE}}$ to view the new range (the default zoom factor is 10 in both x and y directions). Press $\boxed{\text{GRAPH}}$ to view the graph again. Trace to the intersection point and zoom in again to determine the coordinates of a better approximation of this zero. Repeat this process of tracing and zooming in until it is possible to approximate the zero of the function to the nearest thousandth.

Reset the Range to the values in part (4). Graph the function again and then use Trace and the automatic zoom-in to determine the coordinates of the other zeros.

b. Numerically, the zeros may be determined efficiently with the use of a program. To enter a program, follow the steps below carefully:

Press $\boxed{\text{PRGM}}$ then press the right cursor arrow to highlight the word EDIT. Press 2 $\boxed{\text{ENTER}}$ to enter the program into program location 2. The first two lines of the screen will appear as below.

```
Prgm2:
:


```

The name of the program is stored on the first line, immediately following the characters **Prgm2:**. The calculator is in alpha mode for this first line.

Enter the following code, pressing $\boxed{\text{ENTER}}$ after each line of code to move the cursor down to the next line. Keying instructions for each line are listed in the second column. Comments describing the operation performed by each line are provided in the third column.

Code	Keying Instructions	Comments
Prgm2: BOARD	Enter the name BOARD by locating the capital letters above various keys on the calculator.	Title of program.
Disp "X"	Press $\boxed{\text{Prgm}}$, press the right cursor arrow to highlight I/O, then press $\boxed{\text{ENTER}}$. Press $\boxed{\text{2nd}}$ $\boxed{\text{ALPHA}}$, then " $\boxed{\text{X/T}}$ " (Note: " is an alpha-character. It is found above the + sign.)	Displays X on the text screen.
Input X	Press $\boxed{\text{Prgm}}$, highlight I/O, then 2 for Input. Press $\boxed{\text{X/T}}$ for X.	Allows user to input a value for X.
Disp Y_1	Press $\boxed{\text{Prgm}}$, highlight I/O, press $\boxed{\text{ENTER}}$, then press $\boxed{\text{2nd}}$ $\boxed{\text{VARS}}$ (for Y-variables) $\boxed{\text{ENTER}}$ (to print Y_1 on the screen).	Displays the value in memory location Y_1.

Exit the editing mode by pressing $\boxed{\text{2nd}}$ $\boxed{\text{CLEAR}}$ to Quit. To use program BOARD, the function to be evaluated must be entered as function Y_1. To run program BOARD, press $\boxed{\text{Prgm}}$ 2 then press $\boxed{\text{ENTER}}$. The letter X appears on the screen, followed by a question mark on the next line. Enter 30 $\boxed{\text{ENTER}}$ to find the y-coordinate when $x = 30$. Press $\boxed{\text{ENTER}}$ to begin the program again to enter another number. Press 29 $\boxed{\text{ENTER}}$ to find the y-value when $x = 29$. Notice that Y is positive for $x = 30$, and negative for $x = 29$. The zero must be between these two values of x. Continue refining the value of x, to obtain a y value closer to zero each time. Obtain a value for x that is accurate within one-thousandth of the exact value. This method can be used to build a table of values for a given function.

7. Suppose that the board is 10 feet long. How far does its tip sag below the horizontal when you are standing at the end of the board? Repeat for a board that is 12 feet long.

   a. To answer part (7) graphically:

   Use the Trace key to move the blinking cursor to the point whose x-coordinate is as close to 10 as possible. Zoom in and repeat the Trace process. Repeat until the x-coordinate is as close as desired to 10 and then record the corresponding y-coordinate. Will the answer you obtained for y in this manner be less than or greater than the value of y when x is exactly equal to 10? Explain. Repeat this process for the 12-foot board.

b. To answer part (7) numerically, use program BOARD in part (6), or the following:

    i. Enter 4 $\boxed{\times}$ 10 $\boxed{\wedge}$ 3 – 115 $\boxed{\times}$ 10 $\boxed{x^2}$ – 50 $\boxed{\times}$ 10 $\boxed{\text{ENTER}}$. To find the value for $x = 12$, press the up-arrow cursor key. Replace each 10 with 12, and press $\boxed{\text{ENTER}}$. Pressing the up-arrow cursor key allows the last calculation to be *replayed*, also allowing editing of the last expression entered.

    ii. Another numeric method to determine the sag of the board when the function to be evaluated is stored in Y_1 follows:

To store 10 in memory location X, enter 10 then press $\boxed{\text{STO>}}$ $\boxed{\text{XIT}}$. To evaluate Y_1 for this X, press $\boxed{\text{2nd}}$ $\boxed{\text{VARS}}$ (to obtain Y-VARS, the Y variables), then press 1 to obtain the y-variable Y_1. Press $\boxed{\text{ENTER}}$ to display the value of Y_1 for X = 10. To determine Y_1 when X = 12, store 12 in memory location X and repeat as above.

8. How far from the built-in end of the board must you stand for the deflection at that point to be 4 inches? 0.8 inches? Explain.

Press $\boxed{\text{Y=}}$ then enter $Y_2 = -4000$. Set the RANGE as [0, 30] with x-scale of 5, and [–16000, 0] with y-scale of 1000. Press $\boxed{\text{GRAPH}}$ to graph the function Y_1 and Y_2. The horizontal line $y = -4000$ will be graphed. Why might graphing this line be helpful in solving part (8)?

Use Trace and the automatic zoom-in to determine the x-coordinate of the point whose y-coordinate is –4000. When you use the $\boxed{\text{TRACE}}$ key, it will trace along the first function graphed. In what follows, it will be necessary to trace along the graph of $y = 4x^3 - 115x^2 - 50x$. Press the up or down cursor keys to trace along the correct function.

Change Y_2 appropriately to answer the second question.

9. If this function is accurate for diving boards up to 30 feet in length, what length of board will have the greatest horizontal deflection if you are standing at the end of it? What will be the horizontal deflection for this board?

Reset the range as before. Press $\boxed{\text{Y=}}$. To graph only the function Y_1, move the cursor over the highlighted equal sign after Y_2. Press $\boxed{\text{ENTER}}$. The highlighting will be turned off, indicating that the graph option for the horizontal line is no longer active. Graph Y_1 and use trace and zoom to determine the coordinates of the lowest point on the graph.

10. Considering the results of part (9), does this function work for boards of lengths up to 30 feet? Explain.

11. How would this situation change if a smaller or larger person than you were to stand on the diving board?

12. What other questions might you ask about this situation? How might you change this situation to make it more interesting to you?

13. What other situations not involving diving boards might have similar mathematical models? Explain.

B.6 Area

Program AREA assists the user in building intuition for the use of rectangles to approximate the area under the curve. The program sketches the graph of a function $y = f(x)$. If the function is monotonic, AREA draws two approximating rectangles, one showing an underestimate for the curve on the screen, the other showing an overestimate for the curve on the screen. The sizes of these areas can be compared visually. Running the program several times, each time reducing the size of the domain without changing the range, allows the user to view approximating rectangles and compare these to the actual area under the curve of the function over successively smaller intervals for x.

After viewing the graphical information from AREA several times, display the values of the areas of the lower and upper approximating rectangles. Additional code is provided for that purpose.

CODE: AREA

The following code provides a graphical experience in determining the usefulness of using rectangles to approximate the area under the curve of a monotonic function. The graphical code requires 58 bytes of memory.

Code	Comments
All-Off	Turn off Y-variables for graphing.
Disp "Xmin"	Display Xmin.
Input Xmin	Input the value of Xmin.
DrawF Y_1	Sketch the graph of function Y_1.
$Y_1 \to R$	Store the value of Y_1 in R.
Line(X, $\emptyset$, X, R)	Sketch line from (X, $\emptyset$) to (X, R).
Xmin $\to$ X	Store the value of Xmin in X.
$Y_1 \to L$	Store the value of Y_1 in L.
Line(X, $\emptyset$, X, L)	Sketch line from (X, $\emptyset$) to (X, L).

(continued)

(continued)

Code	Comments
Pause	
DrawF L	Sketch line $y = L$.
Pause	
Draw F R	Sketch line $y = R$.
Pause	
DispHome	Return to the home screen.

For a function f that is monotonic, either always increasing or always decreasing over the domain chosen for the viewing rectangle, adding the following code allows you to view the values of the areas of the lower and upper approximating rectangles. The values of the areas are displayed as left (L.AREA) and right (R.AREA) areas. The user determines which of these is lower and which is upper. Program AREA also computes and displays the difference in these areas. The code below can be entered at the end of the code above without any changes being made to either. The full code requires 115 bytes of memory.

Code	Comments
Xmax–Xmin $\to$ H	Determine the x-interval width.
Disp "L.AREA"	Display L.AREA.
LH $\to$ L	Store the value of the left-endpoint area in L.
Disp L	Display the value of L.
Disp "R.AREA"	Display R.AREA.
RH $\to$ R	Store the value of the right-endpoint area in R.
Disp R	Display the value of R.
R–L	Determine the difference between R and L area.
Disp "DIFF"	Display DIFF.
Disp Ans	Display the value of R–L.

OPERATION

Preparation

Enter the formula for the function to be graphed into function Y_1. Set the Range to display the graph of the function on an interval over which it is monotonic. Choose the complete range for the domain chosen. For illustration purposes, it is best to choose an interval for which $f(\text{Xmax}) \neq 0$.

Input

Enter the value for Xmin at the prompt. Press the $\boxed{\text{ENTER}}$ key to continue.

Output

The graph of f is sketched on the interval set by the user, with the value of Xmin chosen from the program. Vertical lines are sketched from the x-axis to the graph of f. Press enter to sketch the horizontal lines $y = f(\text{Xmin})$ and $y = f(\text{Xmax})$. These lines correspond with $y = \text{Ymin}$ and $y = \text{Ymax}$ for monotonic functions f.

To build intuition for using rectangles to estimate the area under the curve, run AREA several times. Each time use a value of Xmin that is larger than the previous choice, but smaller than Xmax. When the difference between Xmin and Xmax is small enough, the graph of f is imperceptible from the graphs of $y = f(\text{Xmin})$ and $y = f(\text{Xmax})$.

After using the first collection of code to investigate under- and overestimates of area using approximating rectangular regions, enter the rest of the code into program AREA. In addition to the output described, the amended program displays the area of the lower approximating rectangle, the upper approximating sum, and the difference in these values.

Sample Run

To illustrate the use of AREA, enter the formula sin X into Y_1. Set the Range to graph $f(x) = \sin x$ on the viewing rectangle [0, 1.57] by [0, 1.5]. Run program AREA several times, for Xmin equal to 0, .5, 1, 1.25, and 1.5. Enter the second set of code above and run AREA again with the same choices of Xmin. Output for Xmin = .5 is as follows: L.AREA = .5129853263, R.AREA = 1.069999661, and DIFF = .5570143344.

B.7 Behavior

Program BEHAVIOR allows the user to investigate the behavior of a function near a given point or to investigate the end behavior (both left and right) of a given function. The program generates a table of values for the function. This program uses 51 bytes of memory.

CODE

Behavior Near a Point

Code	Comments
Disp "A"	Prompt for input.
Input A	Input initial value for A.
Disp "H"	Prompt for input.
Input H	Input increment.
Lbl 1	First label to Goto.
A + H → X	Store A + H in X.
Disp "X"	Display X.
Disp X	Display value of X.
Disp "Y$_1$"	Display Y$_1$.
Disp Y$_1$	Display value of Y$_1$.
Pause	Stop action temporarily.
H/2 → H	Divide value of H by 2.
Goto 1	Repeat from Lbl 1.

End Behavior

Change the second to the last line in the code given above from

$$H/2 \to H \text{ to } H \times 2 \to H$$

OPERATION

Preparation

To use this program, store the formula for the function to be used in Y$_1$ in the $\boxed{Y=}$ menu. For example, if the function is $f(x) = x^2 - 10$, then function Y$_1$ must be entered as $X^2 - 10$.

Input

The input for this program depends on whether it is being used to investigate the behavior of a function near a given point or to investigate the end behavior of a function.

To investigate the behavior of a function near a point, enter the x-coordinate of that point at the **A?** prompt. Enter the value of an increment to be used at the **H?**

prompt. The program evaluates the function for X = A + H. It then divides the value of H by 2 and repeats the process. To investigate the behavior of the function for values of *x* to the right of A, input a small (such as 0.5) positive value for H. Enter a small negative value for H (such as –0.5) to investigate the behavior of the function for values of *x* to the left of A.

To investigate the end behavior of a function, the input process is essentially the same. The only difference is that the program multiplies the value of H by 2 (instead of dividing it by 2) before repeating the process. To investigate the right end behavior of a function, input a large (such as 100) positive value of H, and for the left end behavior, input a large (such as –100) negative value of H. To simplify table building, choose A to be zero when investigating end behavior.

Output

The program outputs successive values of *x* and *y* for the function stored in function Y_1. The first value of *x* is A + H (A and H are entered by the user). The succeeding values of *x* depend on how the program is being used. For behavior near a point H is divided by 2. For end behavior H is multiplied by 2. At each pause in the program, press the $\boxed{\text{ENTER}}$ key to continue. The program does not terminate.

To terminate the program, press the $\boxed{\text{ON}}$ key and select Option 2: Quit.

Sample Run

To investigate the behavior of the function $f(x) = \dfrac{(\sin x)}{x}$ for *x* just smaller than zero, make sure you entered the second to the last line of the code properly and then enter function Y_1 as

$$\sin X / X$$

Run BEHAVIOR using A = 0 and H = –0.5. After each pause in the program, press $\boxed{\text{ENTER}}$. A table of values for the function is generated using *x*-values of –.5, –.25, –.125, –.0625, etc. Corresponding *y*-values generated are .9588510772, .989615837, .9973978671, and .9993490855, respectively. To terminate the program, press the $\boxed{\text{ON}}$ key and choose 2 to Quit.

Adaptation

BEHAVIOR can easily be adapted to print out values for up to four functions. To display *y*-values for function Y_2 as well as function Y_1, insert the following lines of code into program BEHAVIOR immediately following the line **Disp Y_1**:

Disp "Y_2"

Disp Y_2

To display additional function values, insert the lines as above, replacing Y_2 with the function whose values are to be displayed.

B.8 Conics

Program CONICS sketches the graph of the general quadratic whose equation is

$$Ax^2 + Bxy + Cy^2 + Dx + Ey + F = 0$$

where the parameters A, B, C, D, E, and F are chosen by the user where $C \neq 0$. To assure that circles appear as circles and ellipses appear in the proper dimensions, once you have entered the Range, use the option SQUARE found in the $\boxed{\text{ZOOM}}$ menu. CONICS requires 137 bytes of memory.

CODE: CONICS

Code	Comments
All-Off	Turn off Y-variables for graphing.
Lbl 1	First place to Goto.
Disp "A"	Display the letter A.
Input A	Input value of A, store in A.
Disp "B"	Display the letter B.
Input B	Input value of B, store in B.
Disp "C"	Display the letter C.
Input C	Input value of C, store in C.
Disp "D"	Display the letter D.
Input D	Input value of D, store in D.
Disp "E"	Display the letter E.
Input E	Input value of E, store in E.
Disp "F"	Display the letter F.
Input F	Input value of F, store in F.
DrawF (– (BX + E) + $\sqrt{((BX + E)^2 - 4C\,(AX^2 + DX + F)))}$ / (2C)	Graph half of the conic section.
Draw F = (– (BX + E) – $\sqrt{((BX + E)^2 - 4C\,(AX^2 + DX + F)))}$ / (2C)	Graph the other half of the conic section.
Pause	
Goto 1	Return to Lbl 1 and repeat.

OPERATION

Preparation

Write the equation of the conic to be sketched in its general quadratic form as shown on page 233, and determine the values of the parameters A, B, C, D, E, and F where $C \neq 0$. Set the Range to graph the conic section on a suitable viewing rectangle. If you want a viewing rectangle that uses the same x and y scale, first set the Range, then use the option SQUARE found in the $\boxed{\text{ZOOM}}$ menu. Note: If $C = 0$, solve the general quadratic for y and graph the conic directly.

Input

CONICS asks for the values of A, B, C, D, E, and F in the general quadratic equation, $C \neq 0$, $Ax^2 + Bxy + Cy^2 + Dx + Ey + F = 0$. Enter these values at the appropriate prompts.

Output

CONICS displays the graph of the equation entered in its general quadratic form. It first displays one half of the conic, and then displays the other half. Each half is a function of y in terms of x.

Sample Run

To graph the ellipse $\dfrac{x^2}{4} + \dfrac{y^2}{9} = 1$, set the Range to sketch the ellipse on the viewing rectangle -4.8, $4.7]$ by $[-3.2, 3.1]$. Run CONICS. At the appropriate prompt, enter the values of $A = .25$, $B = 0$, $C = 1 \div 9$, $D = 0$, $E = 0$, and $F = -1$. CONICS will display the graph of the ellipse.

B.9 Family

The program FAMILY sketches the graphs of members of a family of functions. Family members sketched are related to the function $y = f(x)$ by the parameters A, B, C, and D in the equation

$$y = A \cdot f(Bx + C) + D$$

where f is a function and A, B, C, and D are real numbers. This program uses 36 bytes of memory.

CODE: FAMILY

Code	Comments
All-Off	Turn off all Y-variables for graphing.
Disp "A"	Prompt for input.
Input A	Input A.
Disp "B"	Prompt for input.
Input B	Input B.
Disp "C"	Prompt for input.
Input C	Input C.
Disp "D"	Prompt for input.
Input D	Input D.
DrawF Y_1	Draw graph of $Y_1 = A \cdot f(BX + C) + D$.

OPERATION

Preparation

Enter the formula for the family of functions to be graphed into the $\boxed{Y=}$ menu as function Y_1. For example, to sketch various members of the family of functions defined by $f(x) = A \cdot \sin(Bx + C) + D$, enter Y_1 as

$$A \sin (BX + C) + D.$$

Set the Range to display the family of function whose graphs are to be sketched.

Input

When this program is run, input values for the parameters A, B, C, and D at each corresponding prompt.

Note: This program can be used for families of functions defined by fewer than four parameters. For example, if the family of functions is defined by an equation containing two parameters, store the formula in function Y_1, use A and B for the parameters and run FAMILY. The user is asked to input values for C and D (as well as for A and B). In this case, it does not matter what values are entered for C and D because these are not used in the formula in function Y_1.

Output

After the values for A, B, C, and D are input, the program sketches the graph of the member of the function family defined by the values of the entered parameters. After the graph is drawn, the program stops. Pressing the $\boxed{\text{CLEAR}}$ key restores the home screen. To repeat the program, press the |ENTER| key again. By repeating the program, you may graph several members of the family of functions on the same screen.

Sample Run

To use the program FAMILY to sketch the graphs of members of the family of functions defined by $f(x) = A \cdot \sin(Bx + C) + D$, enter function Y_1 as

$$A \sin(BX + C) + D.$$

Set the Range to [–5, 5] by [–5, 5] with a scale factor of 1 along each axis.

Run FAMILY using A = 1, B = 1, C = 0, and D = 0. A sketch of the graph of $y = \sin x$ is displayed. Press the |CLEAR| key followed by the |ENTER| key to run the program again. Overlay the graph of $y = 3\sin x$ by entering A = 3, B = 1, C = 0, and D = 0. Sketch the graph of $y = 3 \cdot \sin(x + 2) + 1$, by pressing |CLEAR| then |ENTER| to begin the program again and using A = 3, B = 1, C = 2, and D = 1. Restore the home screen by pressing |CLEAR|.

B.10 Integral

The program INTEGRAL outputs various approximations for the definite integral of a function f over the interval [a, b]. It lists the midpoint approximation of a Riemann sum with n subintervals, the trapezoid approximation using n subintervals, and Simpson's approximation for the definite integral using 2n subintervals. The value of n is supplied by the user. This program uses 171 bytes of memory.

CODE: INTEGRAL

Code	Comments
Disp "A"	Prompt for input.
Input A	Input left endpoint of interval.
Disp "B"	Prompt for input.
Input B	Input right endpoint of interval.
Lbl 1	First label to Goto.
Disp "N"	Prompt for input.

(continued)

Code	Comments
Input N	Input number of subintervals.
.5(B − A)/N → H	Determine 1/2 length of subintervals.
$\emptyset$ → M	Initialize midpoint approximation M.
$\emptyset$ → K	Initialize counter for loop.
Lbl 2	Second label to Goto.
A + (2K + 1)H → X	Store value of midpoint of Kth subinterval in X.
M + 2HY$_1$ →M	Add 2HY$_1$ to M.
IS>(K, N − 1)	Add 1 to K, skip next statement if K = N.
Goto 2	Repeat from Lbl 2.
$\emptyset$ → T	Initialize trapezoid approximation T.
$\emptyset$ → K	Initialize counter for loop.
Lbl 3	Third label to Goto.
A + 2KH → X	Store value of left endpoint in X.
T + HY$_1$ →T	Add HY$_1$ to T.
X + 2H → X	Store value of right endpoint in X.
T + HY$_1$ →T	Add HY$_1$ to T.
IS>(K, N − 1)	Add 1 to K, skip next statement if K = N.
Goto 3	Repeat from Lbl 3.
Disp "M"	Display M.
Disp M	Display midpoint approximation.
Disp "T"	Display T.
Disp T	Display trapezoid approximation.
(2M + T)/3→S	Calculate Simpson's approximation.
Disp "S"	Display S.
Disp S	Display Simpson's approximation.
Pause	
Goto 1	Goto Lbl 1 and repeat with new N.

OPERATION

Preparation

Enter the formula for the function to be numerically integrated into Y_1. For example, to determine approximations for definite integrals of the function $f(x) = \sin(x^2)$, function Y_1 must be entered as $\sin(X^2)$.

Input

When you run the program, input the following:

Prompt	Input
A?	Lower limit for the definite integral
B?	Upper limit for the definite integral
N?	Number of subintervals for the interval [A, B]

Output

The program outputs various approximations for the definite integral in the following order:

M: The midpoint approximation using a Riemann sum with N subintervals
T: The trapezoid approximation using N subintervals
S: Simpson's approximation using 2N subintervals

Once M and T are determined, S is calculated by the formula $S = \dfrac{2M + T}{3}$.

The program does not terminate. Press the $\boxed{\text{ON}}$ key to terminate INTEGRAL.

Sample Run

To use this program to approximate $\int_0^1 \sin(x^2)\,dx$ using N = 10 subintervals, enter function Y_1 as $\sin(X^2)$. Run program INTEGRAL. Input A = 0, B = 1, and N = 10. The following results should be obtained:

M:	Midpoint approximation (10 subintervals):	0.3098162946
T:	Trapezoid approximation (10 subintervals):	0.3111708112
S:	Simpson's approximation (20 subintervals):	0.3102678001

B.11 Inverse

The program INVERSE sketches the graph of a function $y = f(x)$ on the viewing rectangle [1.5Ymin, 1.5Ymax] by [Ymin, Ymax] where Ymin and Ymax are chosen by the user. INVERSE then plots the graph of the inverse relation for a function. The viewing rectangle set by INVERSE uses the same scale for x and y. The program uses 91 bytes of memory.

CODE: INVERSE

Code	Comments
ClrDraw	Clear the graphics screen.
All-Off	Turn off Y-variables for graphing.
Disp "Ymin"	Prompt for input for Ymin.
Input Ymin	Input left endpoint of interval, store in A.
Disp "Ymax"	Prompt for input for Ymax.
Input Ymax	Input right endpoint of interval.
(Ymax–Ymin)/64→Q	Determine step size and store value in Q.
Int(10Q) → Yscl	Determine tick marks, store value in Yscl.
1.5Ymin → Xmin	Set Xmin value.
1.5Ymax → Xmax	Set Xmax value.
Yscl → Xscl	Set Xscl value.
DrawF Y_1	Sketch the graph of function in Y_1.
Ymin → X	Store the value of A in X.
Lbl 1	First label for Goto.
PT-On (Y_1, X)	Find $f(X)$ and plot $(f(X), X)$.
X + Q → X	Store X + Q in X.
If X < Ymax	Compare X to Ymax.
Goto 1	If X < Ymax, repeat from Lbl 1. If not, end.

OPERATION

Preparation

Enter the formula for the function $y = f(x)$ into the $\boxed{Y=}$ menu into function Y_1. For example, to plot the graph of the inverse relation for the function $f(x) = x^2$, function Y_1 must be entered as X^2. Program INVERSE superimposes the graph of the inverse relation on the graph of the function. Choose Range settings for y to provide an appropriate viewing rectangle for the graph of the function and its inverse relation.

Input

When you run the program, input the following:

Prompt	Input
Ymin?	The left endpoint of the interval for the range of the function and the domain of its inverse

Ymax? The right endpoint of the interval for the range of the function and the domain of its inverse

Choose Ymin to be less than Ymax. The program terminates with an error message if Ymin is chosen to be greater than Ymax.

Output

The program will sketch the graph of the function $y = f(x)$ stored in Y_1 and plot the graph of the inverse relation $(f(x), x)$ for this function on a viewing rectangle whose x and y scales are the same. The graphs of the function and its inverse relation are displayed on the screen when the program terminates. To run the program again, press the $\boxed{\text{CLEAR}}$ key to return to the home screen. Press $\boxed{\text{ENTER}}$ to start the program again.

Sample Run

To sketch a graph of the function $f(x) = x^2$ and its inverse relation on the viewing rectangle [–4.5, 4.5] by [–3, 3], enter X^2 into function Y_1. Run program INVERSE and input Ymin = –3, Ymax = 3. This results in the display of the graphs of the parabolas $y = x^2$ and $x = y^2$ on the screen.

Modification

To show the reflective properties of inverses, it might be helpful to sketch the line $y = x$. This may be done from the program by inserting the line **DrawF X** at the end of the program, or it may be done interactively after the program has terminated.

B.12 Newton's Method

NEWTON'S METHOD-GRAPHICAL

Program NEWTON.G uses a modification of Newton's Method to approximate a zero of a function f. If a is the first approximation of a zero of f, Newton's Method obtains the next approximation as the x-intercept of the line tangent to the graph of $y = f(x)$ at the point $(a, f(a))$. This next approximation b is obtained by the formula

$$b = a - \frac{f(a)}{f'(a)}.$$

The modification of Newton's Method is that a numerical approximation is used for the value of the derivative $f'(a)$. The approximation is

$$f'(a) \approx \frac{f(a + h) - f(a - h)}{2h}$$

where $h = 10^{-6}$. This approximation for the derivative is obtained with the NDeriv function of the Texas Instruments calculator. This program uses 119 bytes of memory.

CODE: NEWTON.G

Code	Comments
ClrDraw	Clear the graphics screen.
All-Off	Turn off Y-variables for graphing.
DrawF Y_1	Sketch the graph of the function in Y_1.
Pause	
Disp "A"	Prompt for input.
Input A	Input initial approximation.
Lbl 1	First label to Goto.
A → X	Store A in X.
NDeriv(Y_1, E – 6) → M	Store $f'(A)$ in M.
Y_1 → F	Store $f(A)$ in F.
Disp "F(A)"	Display F(A).
Disp F	Display the value of $f(A)$.
Pause	
Line(A, F, A, ∅)	Sketch line from $(A, f(A))$ to $(A, ∅)$.
DrawF M(X – A) + F	Graph tangent line at $(A, f(A))$.
Pause	
A – F/M → R	Calculate next approximation, store in R.
If abs(A – R) < E–3	Compare successive approximations.
Goto 2	If close, Goto end of program.
Disp "A"	Display A.
Disp R	Display next approximation.
R → A	Store R in A.
Goto 1	Repeat with next approximation.
Lbl 2	End of program label.
Disp "R="	Display R.
Disp R	Display approximate root.

OPERATION

Preparation

To use this program, store the formula for the function in function Y_1. For example, if the function is $f(x) = x^2 - 10$, enter function Y_1 as $X^2 - 10$. Set the Range to provide an appropriate viewing rectangle that displays a root of the function.

Input

When the program is run, the graph of the function stored in Y_1 is displayed. If the graph does not show a root for the function stored in Y_1, terminate the program, reset the Range, and run the program again. Press $\boxed{\text{ENTER}}$ to input the initial approximation for the zero of the function at the first prompt **A?**. The initial approximation is stored in A.

Output

The program first sketches the graph of the function stored in Y_1. It outputs values of A and $f(A)$. NEWTON.G then sketches the graph of a vertical line from the x-intercept A of the previous tangent to the graph of f. The tangent line is drawn to intersect f at $(A, f(A))$. After each output line or graph, press $\boxed{\text{ENTER}}$ to continue. The program will terminate when the absolute value of the difference between two successive approximations is less than 10^{-3}.

Sample Run

To use the program NEWTON.G to find a positive zero of $f(x) = x^2 - 10$, first enter $X^2 - 10$ into function Y_1. Set the Range to graph f on the viewing rectangle $[0, 6]$ by $[-10, 21]$.

Run NEWTON.G using A = 1 as the initial approximation. (This allows a few separate tangent lines to be shown.) Successive values of A and $f(A)$ are displayed, followed by the graphs of a vertical line through the x-intercept of the tangent line to the graph of f and the tangent line through $(A, f(A))$. When the program terminates, it shows an approximate root of R = 3.162277665.

Note: It is possible to convert this program to the program NEWTON, which does not show the graphical interpretation of Newton's Method.

NEWTON'S METHOD

Program NEWTON is a modification of program NEWTON.G. The lines of code that provide a graphical illustration of Newton's Method are deleted. NEWTON provides a numerical approximation for a zero of a function $y = f(x)$. If $x = a$ is the first approximation of a zero of f, NEWTON obtains the next approximation as the *x*-intercept of the line tangent to the graph of $y = f(x)$ at the point $(a, f(a))$. This program uses 59 bytes of memory.

CODE: NEWTON

The code for this program can be obtained by adding one line, deleting several lines, and changing four lines of the code for the program NEWTON.G. The final code requires 59 bytes of memory.

Code	Comments
Lbl 1	First label for Goto.
NDeriv(Y_1, E – 6) → M	Store $f'(A)$ in M.
X – Y_1/M → R	Calculate next approximation store in R.
If abs (X – R) < E–10	Compare successive approximations.
Goto 2	If close, go to end of program.
Disp R	Display next approximation.
Pause	
R → X	Store R in X.
Goto 1	Repeat with next approximation.
Lbl 2	End of program label.
Disp "R"	Display R.
Disp R	Display approximate root.

OPERATION

Preparation

As with NEWTON.G, the formula for the function must first be stored in function Y_1. It is also necessary to sketch an appropriate graph of $y = f(x)$ on the calculator before running the program.

Input

Before the program is run, input the initial approximation for the zero of the function from the graph of the function by using the trace option of the calculator. Sketch a graph of the function that shows the zero of the function as an *x*-intercept of the graph. Then use the trace function to trace along the graph until the blinking cursor is close to the desired *x*-intercept. The *x*-coordinate of the point obtained from the trace will serve as the initial approximation for the zero of the function. Run NEWTON.

Output

The program displays successive approximations of the root of the equation determined by Newton's Method. The program runs until the absolute value of the difference between two successive approximations is less than 10^{-10}. The program outputs the approximation for the zero as R.

To run the program again to approximate another zero of the same function, use the trace key as before to obtain an initial approximation for the other zero. It is also possible to enter an initial approximation for the root directly before the program is run. For example, if a first estimate for the root is 2, before running NEWTON, enter $2 \rightarrow X$, then run the program.

Sample Run

To use the program NEWTON to find a positive zero of $f(x) = x^2 - 10$, first enter $X^2 - 10$ into function Y_1. Next, graph $y = x^2 - 10$ on [–5, 5] by [–10, 5]. Then use the trace option to trace close to the x-intercept between 3 and 4. Finally, run the program NEWTON. The output should be R = 3.16227866.

ALTERNATE VERSION

Program NEWTON.A is an adaptation of program NEWTON. NEWTON.A uses Newton's Method to approximate a zero of a function f when the value of the initial approximation is entered numerically rather than from a graph of the function. It also uses the formula for the first derivative of f rather than a numerical approximation for the value of $f'(a)$, allowing the approximation for the root to be more accurate. This program uses 61 bytes of memory.

CODE: NEWTON.A

Code	Comments
Disp "A"	Prompt for input.
Input X	Input initial approximation.
Lbl 1	First label for Goto.
$X - Y_1/Y_2 \rightarrow R$	Compute next approximation, store in R.
If abs $(X - R) <$ E–10	Compare successive approximations.
Goto 2	If close, go to end of program.

(continued)

Code	Comments
R → X	Copy R to X.
Disp R	Display next approximation.
Pause	
Goto 1	Repeat with next approximation.
Lbl 2	End of program label.
Disp "R"	Display R.
Disp R	Display approximate root.

OPERATION

Preparation

To use this program, store the formula for the function in function Y_1 and the formula for the derivative of the function in function Y_2. For example, if the function is $f(x) = x^2 - 10$, enter function Y_1 as $Y_1 = X^2 - 10$ and function Y_2 $Y_2 = 2X$.

Input

When the program is run, input the initial approximation for the zero of the function at the prompt **A?**. The initial approximation is stored in X.

Output

The program displays successive approximations of the root of the equation determined by Newton's Method. NEWTON.A runs until the absolute value of the difference between two successive approximations is less than 10^{-10}. The program outputs the approximation for the zero as R.

Sample Run

To use the program NEWTON.A to find a positive zero of $f(x) = x^2 - 10$, first enter $X^2 - 10$ into function Y_1 and $2X$ into function Y_2. Run this program using $A = 3$ as the initial approximation. Notice that successive approximations of the root are displayed. The approximate root is R = 3.16227766.

NEWTON'S METHOD (Using the [Ans] key)

You can get the value of the last computation by pressing the [Ans] key. For example, calculate $5 + 2$ on your calculator. Then press

$$\boxed{\text{Ans}}\ \boxed{+}\ 8\ \boxed{\text{ENTER}}$$

The number 15 should appear (7 + 8 = 15).

This makes the $\boxed{\text{Ans}}$ key very useful in any recursive computation such as Newton's Method. If x_n is an approximation for the zero of a function, then the next approximation is obtained from x_n according to the formula

$$x_{n+1} = x_n - \frac{f(x_n)}{f'(x_n)}$$

To perform Newton's Method directly on the TI-81, enter the initial approximation to the zero of the function. Press $\boxed{\text{ENTER}}$ followed by the code for the function $y = f(x)$:

$$\text{Ans} - f(\text{Ans})/f'(\text{Ans})$$

Pressing the $\boxed{\text{ENTER}}$ key repeatedly will produce successive approximations to the zero of the function.

It is necessary to enter the formula for the function, $f(\text{Ans})$, and the formula for the derivative of the function $f'(\text{Ans})$. For example, to approximate the positive zero of the function $f(x) = x^2 - 10$ using 3 as the initial approximation, press

$$3\ \boxed{\text{ENTER}}$$

and then enter the following code

$$\text{Ans} - (\text{Ans}^2 - 10)/(2\text{Ans})$$

Press the $\boxed{\text{ENTER}}$ key several times. When two successive approximations are identical on the calculator, you are finished. In this case, you should get 3.16227766 as the approximate zero of this function.

B.13 Polar Families

Program POLAR allows the simultaneous display of several graphs of the family of functions $r = A f(B\theta - \frac{C\pi}{4}) + D$ where A, B, C, and D are chosen by the user.

With POLAR, the T-interval is set from Tmin to $E\pi$ where Tmin and E are chosen by the user. POLAR requires 113 bytes of memory.

CODE: POLAR

Code	Comments
ClrDraw	Clear the graphics screen.
All-Off	Turn off all Y-variables for graphing.
Disp "Tmin"	Display Tmin to prompt for input.
Input Tmin	Input Tmin.
Disp "E"	Prompt for input.
Input E	Input E.
$E\pi \to$ Tmax	Store $E\pi$ in Tmax.
Disp "N"	Prompt for input.
Input N	Input N.
(Tmax − Tmin)/N $\to$ Tstep	Determine and store step size for T.
Lbl 1	First Lbl to Goto.
Tmin $\to$ T	Initialize T.
Disp "A"	Prompt for input.
Input A	Input A.
Disp "B"	Prompt for input.
Input B	Input B.
Disp "C"	Prompt for input.
Input C	Input C.
Disp "D"	Prompt for input.
Input D	Input D.
Lbl 2	Second Lbl to Goto.
PT-On $(X_{1T} \cos T, X_{1T} \sin T)$	Plot point (X, Y).
T + Tstep $\to$ T	Increment T.
If T < Tmax	Test to see if T is within T-interval.
Goto 2	Return control to Lbl 2 if T < Tmax.
Pause	If T $\geq$ Tmax, pause to look at graph.
Goto 1	Return to Lbl 1 and repeat.

OPERATION

Preparation

Change the calculator function graphing mode **Param** (for parametric mode) to Store the function $r = A \, f(BT - \frac{C\pi}{4}) + D$ in the function menu under X_{1T}. Set the Range appropriately to graph the family of polar functions. Use the SQUARE option in the ZOOM menu or use a multiple of the default viewing window to make certain that x and y intervals have the same scale. It is important that circles appear as circles and ellipses as ellipses.

Input

At the appropriate prompts, input Tmin, E where $E\pi$ is Tmax, and the number N of points to be plotted. Also input the values of A, B, C, and D to graph the appropriate family member of $r = A \, f(BT - \frac{C\pi}{4}) + D$.

Output

The graph of the polar function determined by the choices of A, B, C, and D is sketched. Pressing ENTER upon completion of the graph runs POLAR again. This allows multiple members of a single polar family to be displayed on the same viewing rectangle.

Sample Run

To graph several members of the family $r = A \, \sin(B\theta - \frac{C\pi}{4}) + D$, change the calculator mode to **Param**, parametric mode. Store the expression A sin(BT – Cπ/4) + D in X_{1T} in the function menu. Set the Range to [–4.8, 4.7] by [–3.2, 3.1]. Run POLAR. At the appropriate prompt, enter Tmin as 0, E as 2, and N as 50. Graph r = sin T by letting A = 1, B = 1, C = 0, and D = 0. Press ENTER to run POLAR again. Graph r = 2 sin T by letting A = 2, B = 1, C = 0, and D = 0. Experiment with other values of A, B, C, and D.

B.14 Range

Program RANGE sets the Range centered at the origin using the same scale settings for both x and y. Using the Range [–4.8F, 4.7F] by [–3.2F, 3.1F], circles, ellipses, and the absolute value function all appear as in textbook examples. RANGE requires 52 bytes of memory.

CODE: RANGE

Code	Comments
Disp "FACTOR"	Display F.
Input F	Input the value of the factor F.
−4.8F → Xmin	Store −4.8F in Xmin.
4.7F → Xmax	Store 4.7F in Xmax.
Int F → Xscl	Store [F] in Xscl.
−3.2F → Ymin	Store −3.2F in Ymin.
3.1F → Ymax	Store 3.1F in Ymax.
Int F → Yscl	Store [F] in Yscl.

OPERATION

Preparation

Store program RANGE in program location 1. This program is used frequently. Storing it in Prgm 1 makes it convenient to use.

Input

Input the value of the factor F that is to be multiplied by the values Xmin, Xmax, Ymin, and Ymax. Choose F > 0 or an error will result when a graph is sketched.

Output

When RANGE is run, the values of Xmin, Xmax, Xscl, Ymin, Ymax, and Yscl are changed to be multiples of −4.8, 4.7, −3.2, and 3.1, respectively. The values of Xscl and Yscl become the greatest integer just less than F.

Sample Run

To set the Range to graph a circle with radius 4 on a viewing rectangle that shows the circle as a circle, run RANGE with F = 2. The calculator Range is displayed as Xmin = −9.6, Xmax = 9.4, Xscl = 2, Ymin = −6.4, Ymax = 6.2, and Yscl = 2. For the functions $f(x) = \sqrt{16 - x^2}$ and $f(x) = |x|$, the screen will appear as at the top of page 250.

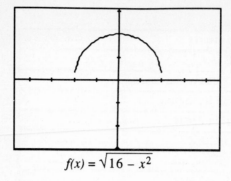

$f(x) = \sqrt{16 - x^2}$

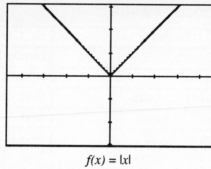

$f(x) = |x|$

B.15 Riemann

The program RIEMANN determines the left endpoint and right endpoint Riemann sum approximations for the definite integral of a function f over an interval [a, b]. If the function f is positive on the interval [a, b], then these Riemann sums will approximate the area under the graph of the function over the interval [a, b]. This program requires 196 bytes of memory.

CODE: RIEMANN

Code	Comments
All-Off	Turn off all y-variables for graphing.
Disp "A"	Display A.
Input A	Input the value of the lower limit of integration.
Disp "B"	Display B.
Input B	Input the value of the upper limit of integration.
Lbl 1	First label to Goto.
Disp "N"	Prompt for input.
Input N	Input number of subintervals.
(B–A)/N → H	Determine length of subintervals.
Disp "L OR R"	Display L for left, R for right.
Input W	Input a letter, store in W.
$\emptyset$ → S	Initialize sum S to zero.
ClrDraw	Clear graphics screen.

(continued)

Code	Comments
DrawF Y_1	Draw graph of Y_1.
If $W \neq L$	Test to determine if right endpoints are wanted.
Goto 2	Send control of program to Lbl 2 if $W \neq L$.
$N - 1 \rightarrow T$	Store the value of $N - 1$ in T.
$\emptyset \rightarrow V$	Initialize V to zero.
$H \rightarrow Z$	Store the value of H in Z.
Lbl 3	Third Lbl to Goto.
$A + VH \rightarrow X$	Determine value of partition point.
$Y_1 \rightarrow Q$	Determine height of rectangle at partition point.
$S + QH \rightarrow S$	Increment sum by area of rectangle.
$X \rightarrow P$	Store the value of X in P.
Line(P, $\emptyset$, P, Q)	Draw part of approximating rectangle.
Line(P, Q, P + Z, Q)	Draw part of approximating rectangle.
Line(P + Z, Q, P + Z, $\emptyset$)	Draw part of approximating rectangle.
IS>(V, T)	Increase V by one. Skip next line if V>T.
Goto 3	Return control of program to Lbl 3.
Pause	
Disp "SUM"	Display the word SUM.
Disp S	Display the value of the sum stored in S.
Pause	
Goto 1	Return control of the program to Lbl 1.
Lbl 2	Second place to Goto.
If $W \neq R$	Test to see if left endpoints are wanted.
Goto 1	Return to Lbl 1 if previous statement is true.
$N \rightarrow T$	Store the value of N in T.
$1 \rightarrow V$	Store one in V.
$-H \rightarrow Z$	Store the value of $-H$ in Z.
Goto 3	Return control of the program to Lbl 3.

OPERATION

Preparation

Store the formula for the function to be used in function Y_1. For example, if the function is $f(x) = x^2$, then enter function Y_1 as X^2. Sketch the graph of the function in Y_1 to determine appropriate Range settings to display a suitable graph.

Input

When you run the program, input the following:

Prompt	Input
A?	Lower limit of integration for the definite integral
B?	Upper limit of integration for the definite integral
N?	Number of subdivisions for the interval
L OR R?	L for rectangles whose height is determined by the function value at the subintervals' left endpoints, or R for rectangles whose height is determined by the function value at the subintervals' right endpoints

CAUTION: When B is chosen to be less than A, choosing L generates rectangles whose height is determined by the function value at the right endpoints and R gives left-endpoint information. If N is chosen to be negative, interval widths are negative, resulting in a change of sign for the value of the sum.

Output

The program first sketches the graph of the function using the Range settings previously set by the user. The interval [A, B] is divided into N equal subintervals. Choosing L results in a sketch of the approximating rectangles for the left endpoint approximation. When the ENTER key is pressed, the value of the left endpoint approximation is displayed. Choosing R results in a sketch of the right endpoint rectangles followed by the value of the right endpoint approximation. Pressing ENTER allows you to repeat the program using the same function and the same interval. Input a new value for N and choose L (left) or R (right) once again.

RIEMANN does not terminate. Press (ON) and choose 2 to Quit to terminate the program.

Sample Run

To use this program to approximate the area of the graph of $f(x) = x^2$ between $x = 0$ and $x = 2$ with Riemann sum approximations, store X^2 in function Y_1. Set the Range settings to

Xmin = -.5	Xmax = 2.5	Scale = 1
Ymin = -1	Ymax = 6	Scale = 1

Run program RIEMANN. To obtain the left endpoint approximation for four subintervals, choose A = 0, B = 2, N = 4, and L. After the rectangles are drawn for the left endpoint approximation, the program will display SUM = 1.75. Press ENTER . Choose N = 4 and R to see the rectangles for the right endpoint approximation. Press ENTER once more to display R = 3.75. After pressing ENTER again, input N = 8. Run RIEMANN to display the graphs and corresponding left and right endpoint approximations. The values for left and right in this case are 2.1875 and 3.1875, respectively.

Notice that since $f(x) = x^2$ is increasing on the interval [0, 2], left endpoints L result in a lower approximating sum for this area, and right endpoints R give an upper approximating sum for this area.

B.16 Secant

The program SECANT draws secant lines for the graph of a function f through the points $(c, f(c))$ and $(c+h, f(c+h))$. It also calculates the slope of each secant line that is drawn. These secant lines can be used to approximate a tangent line to the graph of f at the point $(c, f(c))$. The slopes of the secant lines approximate the slope of the tangent line for small values of h. This program requires 106 bytes of memory.

CODE: SECANT

Code	Comments
All-Off	Turn off all Y-variables for graphing.
ClrDraw	Clear graphic screen.
DrawF Y_1	Draw graph of Y_1.
Pause	
Disp "C"	Prompt for input.
Input C	Input value of C.
Disp "H"	Prompt for input.
Input H	Input increment.
C $\rightarrow$ X	Copy value of C into X.
$Y_1 \rightarrow$ B	Evaluate f(C), store in B.
Lbl 1	First label to Goto.
C + H $\rightarrow$ X	Store the value of C + H in X.
$(Y_1 - B)/H \rightarrow$ M	Calculate and store slope in M.
Disp "M"	Display M.

(continued)

Code	Comments
Disp M	Display value of M.
Pause	
DrawF M(X − C) + B	Draw graph of secant line.
Pause	
H/2 → H	Divide H by 2.
If abs H < .01	Compare |H| to .01.
Goto 2	If |H|<.01 go to end of program.
Disp "H"	Display H.
Disp H	Display new value of H.
Goto 1	Goto 1 and repeat.
Lbl 2	End of program label.
DispHome	Display home screen.

OPERATION

Preparation

Store the formula for the function to be used in function Y_1. For example, to use the function $f(x) = \sin x$, function Y_1 must be stored as sin X. Select an appropriate viewing rectangle for the graph and set the Range.

Input

The program draws secant lines for the graph of the function connecting the points $(C, f(C))$ and $(C + H, f(C + H))$. Press $\boxed{\text{ENTER}}$ after the graph of f has been displayed. Input the value of C and the initial value of H at the prompts.

Output

The program first draws the graph of function Y_1 on the selected viewing rectangle. The value of the slope (M) of the secant line connecting the points $(C, f(C))$ and $(C + H, f(C + H))$ for the values of C and H selected by the user is displayed. Pressing $\boxed{\text{ENTER}}$ once more, SECANT sketches the secant line on the graph of the function. The program replaces H with $\frac{H}{2}$, displays the value of H and repeats the process. Continue pressing $\boxed{\text{ENTER}}$ to see the values of successive slopes M, interval widths H, and resulting secant lines to the graph of f through the points $(C, f(C))$ and $(C + H, f(C + H))$. The procedure is repeated until $|H| < 0.01$, at which time the program ends. With this program, the limit of the slope of the secant lines as the value of H gets closer and closer to zero can be investigated.

Sample Run

To investigate the slope of secant lines for the function $f(x) = \sin x$ at the point $(\frac{\pi}{2}, 1)$, enter function Y_1 as sin X. Using the $\boxed{\text{Range}}$ key, set Xmin = 0, Xmax = 5, Xscl = 1, Ymin = –2, Ymax = 2, Yscl = 1. Run program SECANT with $C = \frac{\pi}{2}$ and H = 2. Press the $\boxed{\text{ENTER}}$ key until four secant lines are drawn. Make note of the values of H and M (the slope of the secant line) as the program progresses. The fourth secant line will have H = .25 and M = –.1243503132. Continue pressing the $\boxed{\text{ENTER}}$ key until the program terminates.

Adaptation

Program SECANT can be used to sketch the graph of a function $y = f(x)$ and a single tangent line to the graph of f at a point $(c, f(c))$. Enter the desired function as described for SECANT. Set the Range. Run SECANT and enter the value of c at the **C?** prompt. Enter H as some small value such as .01. The graph of f is sketched followed by a secant line between two points which are so close together that they coincide on the graphics screen, providing the appearance of the graph of the tangent to f at $(c, f(c))$. Press $\boxed{\text{ON}}$ and choose Option 2 to Quit the program.

B.17 Series

The program SERIES can be used to find a partial sum of a given series. If a_k is the kth term of the series, this program can be used to determine the sum $\displaystyle\sum_{k=1}^{N} a_k$. This program uses 47 bytes of memory.

CODE: SERIES

Code	Comments
Disp "I"	Prompt for input.
Input K	Input initial index value.
Disp "N"	Prompt for input.
Input N	Input final index value.
$\emptyset \to S$	Initialize sum.
Lbl 1	First label for Goto.
$S + Y_1 \to S$	Add term of series to S.

(continued)

Code	Comments
IS > (K, N)	If K > N, skip next line of code.
Goto 1	Repeat from Lbl 1.
Disp "SUM"	Display the word SUM.
Disp S	Display value of partial sum.

OPERATION

Preparation

To use this program, store the formula for the general term of the series in function Y_1 as a function of K. For example, to find partial sums of the series $\sum_{k=1}^{\infty} \frac{1}{k}$, function Y_1 must be entered as 1/K

Input

Once the formula for a_k is stored in function Y_1, then to find the partial sum $\sum_{k=I}^{N} a_k$, input I, the initial value of the index, and N, the final value of the index.

Output

The program will output the value of the partial sum $\sum_{k=I}^{N} a_k$.

Sample Run

For the series $\sum_{k=1}^{\infty} \frac{1}{k}$, enter function Y_1 as 1/K. Run program SERIES and input I = 1 and N = 10. The partial sum is $\sum_{k=1}^{10} \frac{1}{k}$ = 2.928968254. Run the program again using I = 4 and N = 20. The partial sum is $\sum_{k=4}^{20} \frac{1}{k}$ = 1.764406324.

B.18 Spider

Program SPIDER sketches the graph of a distance-versus-time function while displaying the actual movement of the object (spider?) in real time as it climbs vertically on the right side of the screen. The object has the same vertical movement as the y-coordinate on the distance-versus-time graph. The movement of the object is isolated so that the rate at which the object moves is prominent. This program requires 126 bytes of memory.

CODE: SPIDER

Code	Comments
ClrDraw	Clear the graphics screen.
All-Off	Turn off Y-variables for graphing.
(Xmax – Xmin)/95 → D	Determine step size for x, store in D.
(Ymax – Ymin)/63 → Q	Determine step size for y, store in Q.
Xmax – 2D → A	Determine location of spider climbing vertically.
Xmin → X	Store the value of Xmin in X.
Y_1 → B	Store the value of $f(x)$ in B.
Lbl 1	First place to Goto.
PT-On (X,Y_1)	Plot the point (X, Y_1).
PT-On (A,Y_1)	Plot the point (A, Y_1).
PT-On (A – D,Y_1)	Plot the point (A – D, Y_1).
If abs (Y_1 – B) < Q/5	Test size of y movement.
Goto 2	Control moves to Lbl 2 if previous line true.
PT-Off (A, B)	Turn off point (A, B).
PT-Off (A – D, B)	Turn off point (A – D, B).
Lbl 2	Second place to Goto.
If X > (Xmax – 5D)	Test for termination of graph.
Goto 3	Control moves to end if previous line true.
Y_1 → B	Store value of Y_1 in B.
X + D → X	Store value of x + step in X.
Goto 1	Control returns to Lbl 1.
Lbl 3	End of program.

OPERATION

Preparation

Enter the formula for the function to be displayed in function Y_1. Set the Range to display the graph in an appropriate viewing rectangle.

Input

This is no input for this program.

Output

The graph of the function in Y_1 is sketched as an object moves vertically on the right side of the screen. The object has the same vertical movement as points on the graph of f.

Sample Run

To graph $f(x) = \sin x$ and observe the rate at which f changes, store the formula sin X in function Y_1. Set the viewing rectangle to be $[-6.3, 6.3]$ by $[-1.5, 1.5]$. Run program SPIDER and watch the object move on the right side of the screen as the function $y = f(x)$ is being drawn. Notice where the object slows, stops, speeds up, etc.

B.19 Tangent

Program TANGENT sketches the graph of a function $y = f(x)$ and then draws tangent lines whose slopes are approximated by the values of

$$m = \frac{f(x + .01) - f(x - .01)}{.02}$$

for each value of x that is used. In addition, as the tangent lines are drawn, the points (x, m) are plotted, leaving a trace of the graph of the derivative function on the screen. This program uses 103 bytes of memory.

CODE: TANGENT

Code	Comments
All-Off	Turn off all Y-variables for graphing.
ClrDraw	Clear graphics screen.
(Xmax − Xmin)/30 → W	Assign step size to W.

(continued)

Code	Comments
2.5 → C	C determines length of tangents.
DrawF Y_1	Draw graph of function Y_1.
Pause	
Xmin + CW → P	Assign Xmin + C•(step) to P.
Lbl 1	First label to Goto.
P → X	Assign value of P to X.
NDeriv (Y_1, .01) → M	Approximate slope of tangent line.
PT-On (P, M)	Plot point (P, M).
Line (P – CW,Y_1 – CMW, P + CW,Y_1 + CMW)	Draw tangent line.
Pause	
P + W → P	Change P to P + (step).
If P + CW ≤ Xmax	Check for segment still in domain.
Goto 1	If in domain, Goto Lbl 1 and repeat.
Disp "End"	End of program.

OPERATION

Preparation

To draw the graph of the derivative of the function f, store the formula for $y = f(x)$ in function Y_1. Also, an appropriate viewing rectangle must be set using the $\boxed{\text{Range}}$ menu before running this program. It is a good idea to first sketch the graph of the function to insure an appropriate graph is used by the program.

Input

Other than storing the formula for the function, there is no input for this program.

Output

The program first sketches the graph of the function. Each time the $\boxed{\text{ENTER}}$ key is pressed, the program draws the line tangent to the graph at $(x, f(x))$ and it plots the point (x, m) where m is the slope of the tangent line. It repeats this process for about 25 different values of x. When the program is completed, it will display the word End. To view the completed graph, simply press the $\boxed{\text{GRAPH}}$ key.

Sample Run

To sketch the graph of $f(x) = \sin x$ and its derivative, enter function Y_1 as $\sin X$. Set the viewing rectangle to the Trig settings from the ZOOM menu. Run program TANGENT. To draw the tangent lines and plot the points on the graph of the derivative function, press the ENTER key several times. The points on the graph of the derivative function should resemble the graph of the cosine function. When the program is completed, press the GRAPH key to view the graphics screen again.

B.20 Vectors

Programs VECTORS.2D and VECTORS.3D are used to compute the sum, difference, and dot product of two vectors, and to compute the norm of the sum and difference of the vectors. In addition, VECTORS.2D illustrates the sum and difference of vectors, and VECTORS.3D computes the cross product for three-dimensional vectors. The code for VECTORS.2D is written so that it is easily revised to accommodate the code for VECTORS.3D. VECTORS.2D requires 356 bytes of memory. VECTORS.3D requires 360 bytes of memory.

CODE: VECTORS.2D

VECTORS.2D sketches vectors v_1 and v_1, asks the user to determine which operation is to be performed, performs and illustrates the operation, then sketches the resultant vector and displays the norm. VECTORS.2D uses the matrix operations of the TI-81.

Code	Comments
1 → Arow	Row dimension for matrix A.
2 → Acol	Column dimension for matrix A.
[A] → [B]	Store dimensions of matrix A in matrix B.
Lbl 7	Seventh label to Goto.
Disp "V1"	Display vector V1.
1 → N	Initialize N.
Lbl A	Control returns here from Goto A.
Input [A](1, N)	Input first and second components of vector v_1.
IS > (N, 2)	If N > 2, skip next line.

(continued)

Code	Comments
Goto A	Return control to Lbl A.
Disp "V2"	Display vector V2.
1 → N	Initialize N.
Lbl B	Control returns here from Goto B.
Input [B](1, N)	Input first and second components of vector v_2.
IS > (N, 2)	If N > 2, skip next line.
Goto B	Return control to Lbl B.
Lbl 8	Eighth label to Goto.
ClrDraw	Clear graphics screen.
Line (∅, ∅, [A](1, 1), [A](1, 2))	Sketch vector v_1.
Pause	
Line (∅, ∅, [B](1, 1), [B](1, 2))	Sketch vector v_2.
Pause	
∅ → P	Initialize operation variable.
Disp "OPERATION A(1), S(2), D(3)"	Prompt for operation input.
Input P	Store 1, 2, or 3 in P.
If P = 1	If P=1,
Goto 1	Goto 1.
If P = 2	If P=2,
Goto 2	Goto 2.
Disp "DP"	Print DP for dot product.
[A] [B]^T	Compute dot product.
Disp Ans	Display value of dot product.
Pause	
Goto 6	Send control to Lbl 6.
Lbl 1	First label to Goto.
[A] + [B] → [C]	Add vectors v_1 and v_2.
Goto 5	Send control to Lbl 5.
Lbl 2	Second Lbl to Goto.
[A] − [B] → [C]	Subtract vector v_2 from v_1.
Lbl 5	Fifth Lbl to Goto.

(continued)

Code	Comments
Line ([A](1, 1), [A](1, 2), [C](1, 1), [C](1, 2))	Plot vector $v_1 \pm v_2$.
Pause	
Line ($\emptyset$, $\emptyset$, [C](1, 1), [C](1, 2))	Plot resultant vector.
Pause	
Disp "I J"	Display I J
Disp [C]	Display resultant vector.
Disp "NORM"	Print NORM.
[C] [C]$^T \to$ [C]	Determine norm^2.
$\sqrt{}$ [C](1, 1)	Determine norm.
Disp Ans	Display value of norm.
Pause	
Lbl 6	Sixth Lbl to Goto.
Disp "SAME VECTORS Y(1), N(2)"	Prompt for repeat with same or new vectors.
Input Z	Input for decision.
If Z = 2	If different vectors,
Goto 7	Return control of program to Lbl 7.
Goto 8	Otherwise, return to Lbl 8.

OPERATION

Preparation

Prepare the Range to sketch the vectors on a viewing rectangle in which both v_1, v_2 and the resultant vector $v_1 + v_2$ or $v_1 - v_2$ can be shown. If you need scalar multiples of any of the vectors, determine these before entering the vector into the program.

Input

Run program VECTORS.2D. **V1** is printed on the screen. Enter the first component of v_1 after the prompt ?. Enter the second component for v_1 after the second ?. **V2** is printed on the screen. Enter the first component of v_2 after the prompt ?. Enter the second component for v_2 after the second ?. Press $\boxed{\text{ENTER}}$ twice to display vectors v_1 and v_2. Choose the operation **A**ddition (1), **S**ubtraction (2), or **D**ot Product (3). Once the operation has been completed, choose whether or not the SAME Y(1) or N(2) vectors are to be used in another operation.

Output

Once the vectors are entered, a sketch of vectors v_1 and v_2 is displayed. The operation is chosen and the result of the operation is illustrated and displayed. The norm of the resultant vector is also displayed.

Sample Run

To illustrate and display the sum and difference of vectors v_1 = <2, 4> and v_2 = <3, –1>, set the Range to [–9.6, 9.4] by [–6.4, 6.2]. Run program VECTORS.2D. Enter the x and y components of v_1 after the **V1 ?** and **?** prompts respectively. Enter the x and y components of v_2 after the **V2 ?** and **?** prompts respectively. Press ENTER after each input. Press ENTER to sketch v_1 , then once again to sketch v_2 . Choose the desired operation be pressing 1, 2, or 3 for addition, subtraction, or dot product, respectively. After you press 1, v_2 is sketched with its originating point at the end of v_1. Press ENTER to display the sketch of the resultant vector from the origin to the terminating point of $v_1 + v_2$. Press ENTER to display the resultant vector [I, J] = <5, 3> and NORM = 5.830951895. Run the program again choosing 1 for SAME vectors, then 2 for subtraction. Notice the similarities and differences in the geometric representation of the sum and difference of these two vectors. This time [I, J] = <–1, 5> and NORM = 5.099019514. To compute the dot product, press 1 to use the same vectors once more, then choose operation 3. DP = 2.

VECTORS.3D

VECTORS.3D computes sums, differences, norms of sums and differences, and dot products of two- and three-dimensional vectors and cross products of three-dimensional vectors. The code for VECTORS.3D is a revision of the code for VECTORS.2D. Notice that many of the lines of code remain the same. All of the graphic commands have been removed. Code for the cross product calculations have been added. VECTORS.3D requires 360 bytes of memory. A shortened form of VECTORS.3D that can be used to compute the dot product, the cross product, and its norm is provided in the following adaptations.

Code	Comments
Disp "DIM"	
Input Acol	Column dimension for matrix A.
1 → Arow	Row dimension for matrix A.
[A] → [B]	Store dimensions of matrix A in matrix B.
Lbl 7	Seventh label to Goto.
Disp "V1"	Display vector V1.

(continued)

Code	Comments
$1 \rightarrow N$	Initialize N.
Lbl A	Control returns here from Goto A.
Input [A](1, N)	Input first and second components of vector v_1.
IS > (N, Acol)	If N > Acol, skip next line.
Goto A	Return control to Lbl A.
Disp "V2"	Display vector V2.
$1 \rightarrow N$	Initialize N.
Lbl B	Control returns here from Goto B.
Input [B](1, N)	Input first and second components of vector v_2.
IS > (N, Bcol)	If N > Bcol, skip next line.
Goto B	Return control to Lbl B.
Lbl 8	Eighth label to Goto.
$\emptyset \rightarrow P$	Initialize operation variable.
Disp "OPERATION A(1), S(2), D(3), C(4)"	Prompt for operation input.
Input P	Store 1, 2, or 3 in P.
If P = 1	If P=1, add vectors.
Goto 1	Goto Lbl 1.
If P = 2	If P=2, subtract vectors.
Goto 2	Goto Lbl 2.
If P = 4	If P = 4, determine the cross product.
Goto 4	Goto Lbl 4.
Disp "DP"	Print DP for dot product.
[A] [B]^T	Compute dot product.
Disp Ans	Display value of dot product.
Pause	
Goto 6	Send control to Lbl 6.
Lbl 1	First label to Goto.
[A] + [B] $\rightarrow$ [C]	Add vectors v_1 and v_2.
Goto 5	Send control to Lbl 5.
Lbl 2	Second Lbl to Goto.
[A] − [B] $\rightarrow$ [C]	Subtract vector v_2 from v_1.
Goto 5	Send control to Lbl 5.

(continued)

Code	Comments
Lbl 4	Fourth Lbl to Goto.
[A] → [C]	Store dimensions of matrix A in matrix C.
1 → N	Initialize N.
Lbl C	Send control here to compute next term of cross product.
N – 3 Int(N/3) + 1 → R	Determine N + 1 mod 3.
R – 3 Int(R/3) + 1 → S	Determine R + 1 mod 3.
[A](1, R) [B](1, S) – [A](1, S) [B](1, R) → [C](1, N)	Determine cross product component.
IS > (N, Acol)	Is N greater the the dimension of the vectors?
Goto C	If previous line is true, return to Lbl C. Repeat.
Lbl 5	Fifth Lbl to Goto.
Disp "<I J K>"	Display I J
Disp [C]	Display resultant vector.
Disp "NORM"	Print NORM.
[C] [C]^T → [C]	Determine norm^2.
$\sqrt{\ }$ [C](1, 1)	Determine norm.
Disp Ans	Display value of norm.
Pause	
Lbl 6	Sixth Lbl to Goto.
Disp "SAME VECTORS Y(1), N(2)"	Prompt for repeat with same or new vectors.
Input Z	Input for decision.
If Z = 2 Goto 7	If different vectors, return control of program to Lbl 7.
Goto 8	Otherwise, return to Lbl 8.

OPERATION: VECTORS.3D

Preparation

If you need scalar multiples for any of the vectors, determine these before entering the vectors into the program.

Input

Run program VECTORS.3D. **V1** is printed on the screen. Enter the components of v_1 after each successive **?** prompt. **V2** is printed on the screen. Enter the components of v_2 after each successive **?** prompt. Choose the operation **A**ddition (1), **S**ubtraction (2), **D**ot Product (3), or **C**ross product (4). Once the operation has been completed, choose whether or not the SAME Y(1) or N(2) vectors are to be used in another operation.

Output

Once the vectors are entered, depending on the operation chosen, the resultant vector <I, J, K> is displayed by pressing $\boxed{\text{ENTER}}$, or the dot product (indicated by DP) is displayed. The norm of the resultant vector is also displayed for the sum, difference, or cross product vectors.

Sample Run

To display the sum, difference, dot product, and cross product of vectors v_1 = <1, 2, 1> and v_2 = <1, 3, 2>, run program VECTORS.3D. Enter the x, y, and z components of v_1 after the **V1?** and **?** prompts, respectively. Enter the x, y, and z components of v_2 after the **V2?** and **?** prompts, respectively. Press $\boxed{\text{ENTER}}$ after each input. Choose the desired operation by pressing 1, 2, 3, or 4 for **A**ddition, **S**ubtraction, **D**ot product, or **C**ross product respectively. After pressing 1, press $\boxed{\text{ENTER}}$ to display the resultant vector <I, J, K> = <2, 5, 3> and NORM 6.164414003. Run the program again, choosing 1 for same vectors, then 2 for subtraction. This time <I, J, K> = <0, –1, –1> and NORM = 1.414213562. To compute the dot product, press 1 to use the same vectors, then choose operation 3. DP = 9. To compute the cross product, press 1 to use the same vectors once more, then choose operation 4. <I, J, K> = <1, –1, 1> with NORM of 1.732050808.

Adaptation

VECTORS.3D is adapted here to compute only the cross product and the norm of the cross product for three dimensional vectors. V.3D can be used in place of VECTORS.3D. Vector addition, subtraction, and norms of resultants can be computed using the matrix operation built into the TI-81 as follows.

Entering matrices: Enter matrices A and B by pressing the $\boxed{\text{MATRIX}}$ key, highlighting EDIT and pressing 1 for matrix A or 2 for matrix B. Enter the dimensions for A or B as 1 x N (where N is 2, 3, 4, 5, or 6). Enter the values of each component, pressing $\boxed{\text{ENTER}}$ after each entry.

For matrix addition or subtraction: Add the matrices after quitting the matrix menu by entering [A] ± [B]. Note that [A] and [B] are written in blue above the number keys 1 and 2. Press $\boxed{\text{2nd}}$ to access these.

For norm: To obtain the norm of a vector stored in, say, matrix [A], multiply [A] [A]^T then take the square root of any one of the entries in the matrix, eg. $\sqrt{\ }$ [A](1, 1).

For dot product: Enter matrices [A] and [B] as above. Multiply [A] [B]^T. ^{T} stands for transpose. This command is located in the matrix menu as option 6.

The program below allows the cross product of two vectors to be determined once the vectors are entered in matrices [A] and [B]. This version of VECTORS.3D requires 109 bytes of memory.

Code	Comments
[A] → [C]	Store dimensions of matrix A in matrix C.
1 → N	Initialize N.
Lbl C	Send control here to compute next term of cross product.
N − 3 Int(N/3) + 1 → R	Determine N + 1 mod 3.
R − 3 Int(R/3) + 1 → S	Determine R + 1 mod 3.
[A](1, R) [B](1, S) − [A](1, S) [B](1, R) → [C](1, N)	Determine cross product component.
IS > (N, Acol)	Is N greater the the dimension of the vectors?
Goto C	If previous line is true, return to Lbl C. Repeat.
Disp "<I J K>"	Display I J
Disp [C]	Display resultant vector.
Disp "NORM"	Print NORM.
[C] [C]^T → [C]	Determine norm^2.
$\sqrt{\ }$ [C](1,1)	Determine norm.
Disp Ans	Display value of norm.